The Sustainable Well Series

Series Editor Roy Cullimore

The Application of Heat and Chemicals in the Control of Biofouling Events in Wells, George Alford and Roy Cullimore

Water Well Rehabilitation: A Practical Guide to Understanding Well Problems and Solutions, Neil Mansuy

The Sustainable Well Series

WATER WELL REHABILITATION

A Practical Guide to Understanding Well Problems and Solutions

By

Neil Mansuy

Layne GeoSciences, Inc.

A Subsidiary of Layne Christensen Company

LEWIS PUBLISHERS

Boca Raton London New York Washington, D.C.

Library of Congress Cataloging-in-Publication Data

Mansuy, Neil.
   Well water rehabilitation : a practical guide to understanding
well problems and solutions / Neil Mansuy.
      p.   cm. -- (The sustainable wells series)
   Includes bibliographical references and index.
   ISBN 1-56670-387-5 (alk. paper)
   1. Wells--Design and construction. 2. Wells--Maintenance and
repair. 3. Wells--Fouling. 4. Fouling organisms--Control.
   II. Title. III. Series.
   TD405.M36 1998
   628.1'14—dc21   98-44337
                                                                                        CIP

No claim to original U.S. Government works
International Standard Book Number 1-56670-387-5
Library of Congress Card Number 98-44337
Printed in the United States of America 1 2 3 4 5 6 7 8 9 0
Printed on acid-free paper

Preface

This book is a compendium from a recent workshop presented by Layne Geosciences Inc. and organized by the Regina Water Research Institute, held in Regina, Saskatchewan, Canada. The information contained within this monograph has been gathered from many years of theoretical training and research under the expert guidance of Dr. Roy Cullimore while at the Regina Water Research Institute. In addition to the educational background, much of what is presented here are novel concepts and ideas based upon very extensive practical experience on many thousands of wells and water systems.

One of the primary purposes of this monograph is to further advance the knowledge and understanding of well problems and solutions. There is a lack of published information on the topic of well rehabilitation. Some of the concepts and ideas presented here are also contrary to the current literature and the current understanding. This is most evident in the area of the source of "iron related bacteria", many of the "unsafe" bacterial samples, and the "contamination" of wells. Some of these concepts are from observations made by the author during many years of practical experience with water well rehabilitation. Some of the concepts on well hydraulics are presented for their significance to well rehabilitation issues and are currently under further investigation. Many of these concepts require additional research, in order to further understand well problems and solutions.

In the water well industry, it is important to strengthen and enforce well construction codes, designed to protect our most precious resource. In addition to prevention of some contamination of wells taking place, it is also important to begin to accept the fact that wells need periodic appropriate cleaning and preventative maintenance treatments, just like any other part of a water system. The well has always been the part of an entire water system not getting maintenance cleaning, because it is out of sight. The lack of adequate maintenance is also perpetuated by the historical misconception that aquifers are sterile and in order for a well to experience biofouling, it had to be contaminated. In order for water systems to maintain good water quality, it is most important to prevent biofouling of surfaces within water environments.

I would like to thank Dr. Roy Cullimore for his unending support in the public education in this field and also in preparation of this manuscript. Without the efforts put forth by him, this manuscript would have taken much longer to prepare.

Neil Mansuy has approximately 20 years of extensive, worldwide well rehabilitation experience. This includes both academic training and considerable practical experience. Neil holds a M.Sc. degree from the University of Regina, Regina, Saskatchewan, Canada, specializing in iron-related bacteria and causes of well plugging. Much of his academic training was obtained while working seven years at the Regina Water Research Institute under the guidance of Dr. Roy Cullimore. The practical experiences began while working at the Water Institute. Some of this practical experience involved the application of heat or heat in combination with chemicals at sites in the U.S., Canada, and France. After working at the Regina Water Research Institute, he worked as a private consultant and was president of a company selling water treatment equipment.

During the past nine years he has worked with Layne Geosciences Inc., based in Mission Woods, Kansas, as an aquifer and well rehabilitation specialist. During the past nine years Neil has become recognized as a leading authority on well rehabilitation. Neil gives approximately 100 seminars and workshops a year covering all aspects of well problems and solutions. During the past nine years, assessment of well problems and recommending cost-effective solutions has been performed on thousands of wells across the U.S. and around the world. In recent years, he has recommended solutions to hundreds of wells, aquifers and water systems with "unsafe" bacterial results. Neil has the unique combination of extensive multidisciplinary understanding of well problems and the experience to recommend the most cost-effective solutions.

He continues to make strides in further understanding lost capacity and water quality problems, while continually looking for more effective, environmentally sound methods and procedures for removal of deposits from water systems, wells and aquifers. Neil is also the developer of the patented Layne Anoxic Block System™ for controlling biofouling problems on wells.

Acknowledgments

The experience and information in this book would not be possible without all the outstanding people that I have worked with in the Layne Christensen organization. It has been their confidence in me that has allowed the ongoing work towards improving well rehabilitation procedures. It is through the professionalism portrayed by all levels of the Layne Christensen Company that we continue to lead the industry. I am grateful for the guidance given by Dr. Roy Cullimore while working on my undergraduate and graduate degrees. Special thanks are given to Dr. Roy Cullimore, Natalie Ostryzniuk, and Vincent Ostryzniuk of Droycon Bioconcepts Inc. during the preparation of the manuscript for this book.

I would like to dedicate this book to my three children Trevor, Rebecca, and Chantal for the sacrifices they have made during a lifetime of developing knowledge and experience with well rehabilitation.

CONTENTS

INTRODUCTION

Roy said I graced the presence of the Water Institute – it was a lot of fun! I spent seven years altogether at the Institute where I learned a lot. I would like to keep today informal, by that I mean, I want to hear a lot of your comments and questions. I have worked extensively across the U.S. and around the world. With this varied exposure to many situations, I have had experience with many of the problems that can be associated with water wells and their cost-effective solutions.

I will be covering water well rehabilitation today and since many of you have come to hear about the "iron bacteria," much discussion will focus on microbiological issues as well. From extensive experience with hundreds of similar workshops, I have found it necessary to bring together a multidisciplinary focus instead of linear thinking. It is most important to understand some hydrogeology, geology, geochemistry, microbiology, hydrology, and other disciplines in order to understand the mechanisms of lost capacity and water quality problems on wells and water systems. Too much emphasis is placed on one aspect of a well and taking a "snapshot" in time. With a more encompassing approach to the cause of problems, we then begin to understand the important aspects of well rehabilitation.

Layne Geosciences Inc. is a wholly owned subsidiary of Layne Christensen Company in Mission Woods, Kansas. Layne Geosciences is a technical subsidiary of the company with a group of seventeen specialists including hydrologists, geologists, hydrogeologists, geophysicists, geochemists and I am the microbiologist. Much of our work is with our district offices designing water wells, looking for and developing water resources, solving water quality issues and also well rehabilitation. At this time there are approximately fifty district offices within the U.S., and approximately 25 international offices. Layne Christensen Company does work on approximately four thousand wells per year, This work includes: pump repairs and maintenance, pump installation and water well rehabilitation. Thus we have considerable experience, for many different types of well construction, types of formations, and well problems and solutions.

During this workshop, there will be a different approach to the whole topic of understanding well problems. This approach is taken to generate new insight and to start thinking in concepts. The inter-relatedness of lost capacity and water quality problems experienced on wells will be clearly demonstrated. There are some issues, which will be addressed in a manner that may shock some of you. Because of the extensive practical field (real world) experiences, some things that will be said are contrary to a lot of the literature, and these views are expressed as I see them. For example, there is not a lot of literature in the area of well rehabilitation and groundwater microbiology.

Some of the topics that will be focused on are the total coliform problems we commonly experience in water wells. In other words, they can be called "unsafe" bacterial samples. These are commonly caused by natural indigenous total coliforms; they are not due to contamination events taking place in groundwater wells most of the time. I will be discussing a lot about well hydraulics, the source and the cause of problems, the inability to easily get these problems under control, the difficulties we have in solving some of these problems, and the things that need to be done in order to get well problems under control.

WELL CONSTRUCTION

WATER SUPPLY DEVELOPMENT

There are some basic terms that really need to be defined before we proceed. Aquifers may be confined or unconfined. Unconfined aquifers are also referred to as water table aquifers. These aquifers do not contain a confining layer and the static water level inside the well will be at the same level as the level in the aquifer. A confined aquifer contains relatively impermeable layer(s), commonly referred to as an aquitard. An aquitard most commonly is a clay layer that confines the aquifer, which keeps the aquifer under pressure. Aquifers, which are confined, will be under pressure so that the water level inside the well would be higher than that of the water level in the aquifer. Understanding the type of aquifer is important when understanding deposit formation. Confined aquifers can have minerals come out of solution from the release of the pressure on the formation. There are other types of aquifers (i.e., semi-confined or leaky artesian), which will not be discussed during this workshop. For the purposes of this workshop there are two major types of aquifers, the unconfined aquifers, where the top of the saturated zone is at the atmospheric pressure, and the confined aquifers under some pressure. The type of aquifers for this discussion is not for purposes of yield but more for water chemistry and deposit formation.

Consolidated Formations are wells drilled with open-hole completion. These are much easier to rehabilitate. The barriers to redevelopment come from a gravel pack and fines traveling in from the surrounding formation. In a consolidated sandstone, limestone, dolomite or granite you do not have these barriers to redevelopment.

Unconsolidated Formations are composed of a wide range of materials. Most of the unconsolidated formations are composed of sand and gravel of varying sizes and uniformity.

GEOLOGICAL FORMATION

It is important to understand different well components and construction so that we have a good idea of the fundamental issues involved during well rehabilitation. It is important to know because Layne works on wells that are a few feet deep to wells that are up to 4,000 feet deep. There are wells that will produce, for example, from 5,000 gallons per minute (gpm) and some that will produce only a few gallons per minute. We have some aquifers in the U.S., for example, the Edwards aquifer in Texas, that have wells capable of flowing up to approximately 14,000 gpm. That is natural flow. These are very sizable wells. The pump bowls are 24" in diameter and they still have pumps installed because the water does not flow at that rate all of the time. Even though there are many variations of production capabilities, the concepts and the aspects of well rehabilitation are the same.

There are a wide variety of different well construction types. Many wells are completed as open-hole in rock (consolidated) formations. These open hole rock wells do not require a well screen as water flows through fractures in the consolidated formations. Quite a few regions of North America have these types of wells. The most common consolidated formations are found in sandstone, dolomite, limestone and granite with varying degree of fractures. Open hole wells are traditionally easier to rehabilitate. Easier rehabilitation results from the absence of a well screen and/or gravel pack that often create additional barriers to easy and effective well development.

WELL SCREEN

The most common wells are screened with many different types of perforations. It is very important to understand what type of screen is used for construction. In addition to the type of screen, the type and thickness of the gravel pack and the type of surrounding formation in which the well is completed are also important. The type of screen geometry being used determines development procedures. Whether it is a wire wrap (WWS), pipe-based wire wrap (PBWWS), shutter or louver screen, perforated casing or mills knife perforation. Understanding the type of well screen is very important when it comes to well rehabilitation procedures. The different type of perforations in well screens can create barriers to the easy removal of deposits.

TYPES OF SCREEN OPENINGS

There are a number of different configurations for slots within a well screen to allow the water to pass through the screen material. The difficulties in development can be demonstrated with a bridge slot screen. In a bridge slot well screen, sometimes called a "double louver", the openings are on the sides of the punched bridge. This is not a very good screen design for the purposes of rehabilitation, since during development it is not possible to generate any significant energy in the formation due to the deflections created by the double louver.

The form or geometry of the opening in a well screen may also influence the likelihood of mechanical type losses due to blockage and the degree of difficulty in developing. The common type of well screens found include: slotted PVC pipe, slotted steel pipe (torch cut, mills knife, etc.), wire-wrap well screen, pipe-based wire-wrap well screen, louver or shutter screen, bridge slot (double louver), etc. Depending upon the geographical area of the U.S., many screens have shutter openings and literally are punched from the inside to create a louver. If you go into the southwest states of the U.S., approximately 90% of the wells are constructed with a shutter or louver opening. This type of well screen is selected for its superior strength and open area (full flow shutter screen). When the shutters are created in the screen, the collapse strength is actually greater than the original pipe.

Test data have shown that the collapse strength is commonly 50% greater than the pipe from which it is made. The wells completed with these openings are commonly more difficult wells to develop or rehabilitate. This is because the shutter opening is angled at an approximately 45-degree angle downward and so is more difficult to get the energy of development into the surrounding formation due to deflection of this energy. Wells constructed with shutter screens can be developed very effectively but may require a slightly more aggressive approach. The next type of well screen is the wire-wrapped type. Depending upon the geographical area, this is the more common type which has a continuous wire-wrapped around and tack welded to the vertical rods. Here, these perforations (slots) can become plugged, and clearly, when this happens, water is not going to be able to enter the well. A wire-wrapped screen allows for easier development because there is less deflection of the energy (e.g., high pressure jetting, swabbing, etc.) through the screen. Some parts of the U.S. use a pipe-based wire-wrapped screen. Here, there is a central pipe which has holes cut in it and a rod-based wire-wrap screen on the outside of the pipe. These are commonly used for deeper wells where the wells are commonly 1,000' to 2,000' deep. The internal pipe is necessary to provide additional strength for the wire wrap. If the internal pipe was not constructed as part of the well screen, the continuous wire wrap could break away from the vertical rods and destroy the well screen. There are a wide range of well screen types and they do become very important in determining what needs to be done on a well in general for the purposes of well rehabilitation.

Notice that deposition and growths can build up on the outside of the well screen, and one of the techniques that has been used for years has been to use a wire brush down the inside of the well to clean off the deposited material. Even though cleaning the inside of the well is a beneficial step, this is not going to work most of the time by itself, because much of the plugging occurs deeper out in the surrounding formation. It is important to be able to penetrate beyond the well screens to where these deposits are and then remove them. We have to rely on agents that are able to penetrate beyond the well screen; these agents have historically been predominantly chemicals and more recently, liquid carbon dioxide.

CEMENTED GRAVEL PACK

On occasions, a gravel pack can become cemented to the outside of the well screen. In this example, the gravel pack was pulled out of a well 900' deep attached to the well screen. The idea was to put a new well screen and gravel pack inside the old screen as a liner since the old screen had failed. This well was shot a couple of times with Sonar-Jet™. This is a vibratory explosive used to break up mineral encrustation, but in this case, it did not work, and so we pulled the well screen (it took 100 tons of force to remove the well screen with the cemented gravel pack attached to it). Back at the shop, the gravel pack was still firmly cemented! What this example demonstrates is how tough some of these deposits can be. It also tells me that some of the cemented matrix can be extremely hard and we tend to "dabble" too much during well rehabilitation! When we treat a well we are really only "dabbling" at the problems (scratching at the surfaces, literally) unless we take a more aggressive approach.

Often we think that by putting a little bit of chemistry into a well, everything is going to go into solution and, vòila, everything is going to be fine again. It is often believed that the capacity is going to be easily restored. We have to know how hard that material can become cemented and we have to become much more aggressive in our approach to treatments. The more important aspects of well rehabilitation, although the chemicals are important, are the forceful agitation and the physical redevelopment of the well. You cannot "dabble" at well rehabilitation. If we consider cleaning a well with the pump in place, we again cannot get such effective treatment because we cannot get good agitation. Number one rule in well rehabilitation is to pull the pump most of the time. You cannot get good agitation or the necessary disruptive velocities (energy) if the pump is left in-place and the flushing action for the deposits is limited. The effectiveness of treating with the pump in place is also determined by the length of producing zone, type of deposits plugging the well, the type of screen opening, and the amount of material that needs to be removed.

WELL CASING

The casing does not have any perforations and openings and will go through the nonproductive areas of the aquifer. Holes due to corrosion in the upper casing can also be a source of water infiltration into the well. This water infiltration can be from perched aquifers, which may contain total and/or fecal coliforms that can result in "unsafe" bacterial results. Below that, there is the well screen of varying lengths with different types of openings and different types of geometry. Many wells are constructed with Screen—Blank—Screen configurations. These wells are constructed with screen—blank (casing)—screen—blank—screen which relate to the screens being positioned in the productive zones of the aquifer. This can go on for as many as ten to twenty different sections. The blanks are essentially "casing off" those parts of the aquifer that are not productive although there can be cross connection within the gravel pack on the outside of the casing if they are not cemented. The blank sections are positioned depending upon what is found during the drilling process. When you finally have a well screen in place, it may or may not be surrounded by an envelope of gravel. It is very important for purposes of well rehabilitation to know if your well has a gravel pack because much of the limitation to development and the difficulty in removal of plugging deposits exist at the aquifer-formation interface.

SANITARY WELL SEAL

These seals are designed to eliminate the potential risk of surface water infiltration into the well and causing additional concerns. Faulty seals can result in some coliform problems in the well water. An effective seal eliminates the potential for "true" contamination by total and/or fecal coliforms from surface water and wastewater. It is very important to keep a well head sealed from surface contamination and to keep the area around a well clean.

ANNULAR SEAL

It is also necessary to have an annular seal around a well down at least a part of the casing to prevent the movement of surface waters or perched aquifers and wastewaters down the outside of the casing. These seals are generally installed after the screen, casing and gravel pack have been put in. The seal is generally achieved using cement or grout. This is a present day standard for well construction used all of the time to prevent surface water infiltration. The depth of the annular seal will again vary from one part of the country to another. Some of the older wells do not have an annular seal and can be directly contaminated. When dealing with well problems this is one of the first things that's checked, particularly when there are "unsafe" bacterial samples (not only high coliform). When you get high bacterial counts, unacceptable coliforms, or TNTC (too numerous to count) atypical, overgrown conditions, the first thing to look at is the well construction. At the same time, look at any surface water which might be in the vicinity of the site and could infiltrate into the well. These sources would be direct true contamination events taking place.

DRAINAGE AWAY FROM THE WELL HEAD

A properly completed well should have good drainage away from the top, so that all of the surface water runs away from the well head. Many wells will have a pump house, depending upon climate. It is very important to ensure no standing water is around the top of the well to prevent surface infiltration. Most of the wells that have "unsafe" bacterial problems do not show any evidence of surface water infiltration.

WELL SCREEN AND GRAVEL PACK SELECTION

A well screen has to hold the formation back from the inside of the well and allow water to enter into the well. The purpose of the <u>gravel pack</u> is also to retard the movement of material towards the well by forming an "envelope" of gravel around the screen. Many sand and gravel formations do not lend themselves to being naturally developed (refer to section on naturally developed wells). Many wells experiencing sand-pumping problems, do so, from poorly designed wells and improper selection and placement of a gravel pack. To select the type of gravel pack specifically for a well, a test hole is drilled and sieve analysis is performed for grain and particle size distribution of the aquifer material. Based on the size and uniformity of the aquifer material, a gravel pack is selected which will hold most of that formation back. After selection of the proper gravel pack, it is necessary to select the screen slot size based upon what the gravel pack selection has been. Properly designed and constructed wells should not have sand-pumping problems after an adequate development step. There are, however, sand-pumping problems experienced on many wells around the world, from improper selection and placement of a gravel pack envelope. These comments are not made for purposes of well design, but instead, how they influence loss of specific capacity, well development and rehabilitation, and lead to pump wear.

PVC WELLS AND BIOFOULING

When it comes to the installation of PVC wells or PVC screens, I am not a strong advocate. One of the negative aspects of using this type of screen is the lack of open area. There also has to be a little more concern about the work-over on PVC wells. While it is possible to treat wells with PVC screens, preference is for screens that are more robust. In the environmental industry, many of the extraction, recovery and monitor wells are equipped with PVC screens and it is possible to treat these wells successfully. The principles remain the same. You have to determine what the problems are. Next, you have to determine what chemistry is adequate. You can choose to swab those wells, provided you do not get overly aggressive. It is possible to swab and you can surge. Finally, you can develop those wells in the same way as other wells. The principles remain the same.

We have found, for example, that we have been very successful with the use of procedures such as carbon dioxide on PVC wells and on horizontal high-density polyethylene (HDPE) wells. These wells were approximately 700' long with 2" to 6" diameters horizontal HDPE screens that are used for the extraction of contaminated ground water. It is therefore fairly easy to rehabilitate PVC wells. We have been very successful in developing and rehabilitating horizontal wells constructed of stainless steel and HDPE, with the use of gaseous carbon dioxide and development procedures.

One of the advantages of PVC is that it is not easily corroded. If you think that the PVC is immune to iron deposition, that is not true. You may see just as much iron deposition on a PVC pipe or screen as you can on a steel pipe because most of the iron that is being deposited is not a corrosion by-product. Most of this material is water filtration products as opposed to corrosion by-products. For that material to be corrosion by-products there would have to be total destruction of the pipe in order to have that massive amount of material buildup!

The major difference that has been seen between the PVC and carbon or stainless steel is that the deposition tends to be more even. In the steels, there tends to be more tuberculation, and that is because of the establishment of the SRBs and the subsequent buildup of some corrosion by-products from the steel as well. This means that there may be some differences in the manner in which the biofouling manifests itself, but PVC most certainly is not immune to the deposition of iron. There are a lot of people who cannot see this. They simply question "where is the iron coming from if we do not have iron pipe?" This is an oversimplification. Clearly iron can enter in the ground water itself and if the water chemistry and the environment for bacterial enhancement lend themselves, there will be biologically induced iron deposition.

When I was doing my thesis on the factors that influence well plugging, stainless steel versus PVC screens was tested. It was found that in the presence of stainless steel, the well plugged at a faster rate than the PVC wells. The PVC wells were found to plug at a rate on average 63% slower than stainless steel screens under the same environmentally controlled conditions. It must be indicated that this occurs without any loss of the stainless steel. The initial thought is the electrical activity in the presence of the stainless steel may be a better environment for deposition. Because of other disadvantages with PVC screens, the material of choice is still most commonly stainless steel.

DEVELOPING GRAVEL PACK WELLS

You have to take somewhat of a different approach when developing gravel pack wells, because they are, in general, more difficult to develop. The basic difficulty results from additional barriers to the migration of fines from the surrounding formations. The most difficult aspect of developing a screened gravel pack well is achieving adequate velocity and energy of development at the gravel pack aquifer interface. The thicker the gravel pack, the more difficult it can be. That makes it more difficult to develop because it is difficult to "migrate" the fines whether they are from formation material (clays, sands and fine silts) or mineral and/or biological deposits. We have to literally migrate all of this material through the formation, gravel pack and the openings of the well screen before they can be removed from the well.

With more emphasis on development of wells in recent years, tremendous improvements have been made in restoring capacity to wells. Some wells can be very difficult to develop. Some wells that have screen openings that are as small as 0.010" to 0.014" of an inch, commonly found in Louisiana, are almost impossible to develop because of the difficulty in passing the particles through the openings. We have wells that are excellent producers and will produce 3,500 or 4,000 gpm, but within 5 years will lose most of their production. Very fine sand migration becomes packed up against the gravel pack and is referred to as mechanical blockage. The challenge is to "migrate" these materials through the gravel pack and the screen slots. If these wells were plugged with biological and mineral deposits it would still be easier to dissolve and disrupt these deposits, removing them through the small screen openings. In some regions mechanical blockage is the more common cause of lost capacity. As a result, many of these wells will remain permanently packed so that the capacity may never be restored. I believe that mechanical blockage issues on wells have been historically understated, with emphasis placed on bacterial plugging and mineral encrustation. If the migration of fines from the surrounding formation does not pack around the well, it can experience sand pumping problems causing significant pump wear (Figure 1).

Problem: Pump Wear

Figure 1

THICKNESS OF GRAVEL PACK

I deal a lot with wells that range in diameter from 6" to 36" with the most common diameters being 10", 12", 14", 16", 18", 20", 22". Many of these wells have underream holes or boreholes that are often 36", 48" or sometimes up to 72". This creates a gravel envelope that may be as much as 20" thick on each side of the well screen. The thicker the gravel pack, the more difficult development will be because there is an additional barrier to development at the outside of the gravel pack. The maximum development efficiency has only been achieved where there is less than 8" of gravel thickness. When we talk of development, this involves the generation of energy, which moves out into the surrounding formations. This "energy" transfer is obviously in water and the water is the carrier of that energy by whatever method. Jetting is very limited due to very limited penetration of energy into the surrounding formation. Jetting alone also has energy concentrated in the wrong direction. Line swabbing or surge block is fairly effective in development and is the most commonly used procedure. The method of development depends in large part on the potential for blockage to occur associated with the aquifer type. The ability of achieving velocities associated with the cleaning process is clearly a major factor in the effective depth of penetration of the rehabilitation or development techniques. We have recently been using a well rehabilitation and development tool that utilizes a combination of the energy of gaseous carbon dioxide and a recirculating process through the gravel pack. This process has the advantage of penetrating energy into the formation and washing the gravel pack and surrounding formation. This has recently been very successful on water supply and recovery wells.

NATURALLY DEVELOPED WELLS

Depending on the geographical region of North America, screen type and slot size may vary. Most screened wells are gravel packed. Many of the older wells are naturally developed. These wells do not have any gravel pack, so the well screen is right up against the natural formation. These wells are either perforated in place once the pipe is in the aquifer or the perforated pipe is placed in the aquifer but no gravel pack is used. The development of such a well would remove some of the finer parts of that formation. Naturally developed wells are easier to rehabilitate than gravel-packed wells due to the lack of an additional barrier to development. Where you have a nonuniform formation with lots of variation in the sizes from very fine to very coarse, where it may become more prone to rapid "mechanical" type losses, there is the ability of the fine material to migrate through the larger openings and void spaces of the porous media (aquifer). A uniform formation, even though it could consist of very fine sand, is not going to lend itself to mechanical blockage because the fine material would not migrate as easily from high velocity. The materials are not going to move but will stay where they are. Uniformity or nonuniformity of the formation is as important as the grain sizes within that formation.

INITIAL WELL DEVELOPMENT PROCESS

One of the most important steps of well construction and the step that often does not get enough attention is the development process when a well is first drilled. The initial development process is designed to remove the fine material that is in the aquifer and the gravel pack. The process involves achieving high velocity conditions toward the well screen in order to migrate the fine material from the aquifer and gravel pack through the well screen openings. Thus, when the well goes into production, it would not be experiencing any sand pumping problems. The lower velocities during normal production (rather than during development) would lead to a very low or insignificant risk of materials such as sand continuing to move through the pack and screen slots into the well. In other words, the velocities applied during development are much higher than the velocities that would be experienced during normal production operations. Unfortunately that is not always the case, even in properly designed and constructed wells. Because of the variation of where the water is produced in a well, high velocity conditions can still exist and on rare occasions the well may still experience sand pumping problems.

ENTRANCE VELOCITIES

Approach and entrance velocities are very important when designing wells. Many wells are designed to have an entrance velocity less than 0.1 feet per second. This is a theoretical calculation based upon the assumption that every square inch of a well screen is producing exactly the same amount of water. In practice this most likely often does not exist. Therefore even properly designed and constructed wells can still experience mechanical blockage due to the migration of fines from high velocity conditions. There is a different approach to the prevention of that type of problem. This involves over-designing the wells. To do this a lesser extraction rate (e.g., lower pumping rate) may be used to reduce the velocity of water into and through the well and/or a larger diameter screen can be used. While this would appear to solve the problem it is often not enough. You can still get high entrance velocities of the water entering through the screen slots because of variations of the zones in a well where water is being produced. This is commonly referred to as "well hydraulics."

PUMP TEST

Once a well has been constructed, it is normal to perform a pump test. A pump test has to be done properly to determine a safe yield from the well and aquifer. This is usually done to generate a baseline for well efficiency and specific capacity to be determined. When a well is being monitored, it is not sufficient to just determine the rate of production (in gpm) for that well. Doing this may mean that you could loose 60, 70 or even 80% of the well's capacity before you even know it! By measuring the well efficiency and specific capacity of the well, problems in wells may be detected much sooner so that rehabilitation and treatments can be applied when they have a higher percentage of success and the life of the well can be extended.

DISCHARGE CAPACITY AND SPECIFIC CAPACITY

In addition to understanding well construction, for purposes of understanding well rehabilitation, it is also essential to understand some of the common terms extensively used, such as drawdown, specific capacity, etc. A good understanding of specific capacity is most important since this is the measure of a well most commonly used. When we talk about the efficiency of a well or the loss of well efficiency and loss of specific capacity, it is most important to know what the specific capacity is as a measure of the status of a well. Obviously there is real interest in understanding the failure of wells that can be measured through the loss of specific capacity and how long the specific capacity is maintained. This will mean a loss in production and the importance of keeping good records during operation of a well, in order to respond to a problem with a well as quickly as possible. With a better understanding of specific capacity and well efficiency we begin to understand the important aspects in water well rehabilitation and preventative maintenance of a well. It is very important to evaluate the specific capacity of a well with the same gpm (Q) flow rate. This makes comparisons more accurate. When monitoring a well or a wellfield for purposes of maintenance and rehabilitation, a well is best compared to itself and looking for relative changes. These relative changes have to be under similar conditions of flow, static water level, etc. Even monitoring a well with specific capacity, bacterial assessment, and water chemistry is more valuable when determining relative changes in the same well, that can occur with these parameters.

Preventative maintenance of a well has limitations of being able to remove material with the pump in place. For this reason it is essential to respond to a problem as quickly as possible to prevent deterioration of wells from happening. We are going to be addressing the issues of well production and some of the factors causing losses in order to try understand the mechanisms and thus understand what is necessary to prevent some of these things from happening again.

Even though, in my opinion, specific capacity is not able to determine the beginning of deposition in the void spaces, it is still the best sign we have to give an early warning indication for problems in a well. As discussed further in the section on longevity of treatments, there is commonly a lot of excess production capacity that can exist in a well and the surrounding aquifer. This excess production of the porous media in the aquifer allows you to have a lot of deposition occurring, before it starts impacting the flow of water and possibly the specific capacity. Thus the hydraulic conductivity of the aquifer and porous media in a well could be severely reduced by deposits without specific capacity being reduced.

Specific capacity (SC) is calculated as the gallons per minute (Q) per foot of drawdown (dd). Drawdown is the difference between static water level and pumping water level. A well will have a static water level when the well is not being pumped. This is referred to as the standing water level (in nonpumping conditions). When the well is pumped, the water level will now change in response to the withdrawal of water from the well. This will then be a pumping water level when the well is being pumped. This is the drawdown from the static water level (swl) to the pumping water level (pwl):

$$dd = swl - pwl$$

Discharge Capacity is the gallons per minute that is being pumped from the well and is commonly referred to as "Q".

$$SC = Q / dd$$

For example, if you had a specific capacity of 100 gpm/ft of drawdown, it means that you could pump 100 gpm for every foot of drawdown. The importance of this is, if you are only measuring the discharge capacity and not checking drawdown, you could be pumping at 500 gpm for example, but not notice that the drawdown gradually gets larger and larger. The excessive drawdown results from the well becoming plugged up with materials close to the well screens (well losses) in the gravel pack and porous media (formation losses). It is important to determine if the loss of specific capacity is partially caused by dewatering the aquifer. If you do not measure the drawdown but just rely on the Q value (pumping rate in gpm), this loss in efficiency would not be noticed.

Well capacity is not a true measure of production potential that exists in a well. If you are measuring gallons per minute, you are not going to be able to detect the problems early enough. It is most essential to keep good records. You have to know what your discharge rate (Q) in gpm is, the static water level, and the drawdown so that you can detect changes as early as you possibly can.

WELL EFFICIENCY (WE)

It is difficult to get 100% efficiency on a well. Common targets for well efficiency are approximately 80%. What well efficiency is relates to the cone of (water) depression or influence that is created around the well when the water is being pumped. This cone extends downwards on the outside of the casing and remains at a higher level than the water level in the well during the pumping (pwl). The difference between the drawdown point on the outside of the casing and the pumping water level on the inside of the well is used to calculate the efficiency of the well. Actually we will sometimes have a piezometer on the outside of the casing so that it is possible to measure the drawdown point on the outside of the well. If there is no manner to observe this, then a theoretical calculation has to be used based upon the transmissivity (T) values for the porous medium around the well. To calculate the well efficiency, the feet of drawdown outside the well is divided by the feet of drawdown on the inside the well and multiplied by 100 to determine the percentage efficiency:

WE = drawdown outside well/drawdown inside well casing X 100

What is being determined is the well loss and that well loss is either due to a well screen, gravel pack, or formation losses. These losses can be due to either turbulent flow or losses due to plugging deposits. Due to the losses in the gravel pack, formation or screen it would be extremely rare to have a well efficiency of 100% efficiency. A 100% efficient well would have the same water level outside the casing as the inside of the well. In consolidated formations (open-hole wells) much of the water passes through fractures into the well. Commonly, 25% of the formation will consist of fractures in dolomites, limestone, sandstone and granites and here, the WE is not so significant as it is in the alluvial formations with screened and gravel-packed wells. In order to understand the losses in well efficiency, we need to take a look in subsurface environments at the processes taking place.

LOSSES OF STATIC WATER LEVEL

If you have experienced loss of capacity on a well, the static water level must be compared with the original static water level. Let's say that you had determined that the specific capacity had dropped from 100 gpm/ft to 90 gpm/ft and the static water level has also dropped; the loss of capacity may be partially or completely due to dewatering the aquifer. If the aquifer is being dewatered, the yield may be significantly affected because there is less head. It must be kept in mind when rehabilitating wells that if you do not have as high a static water level as originally, you cannot expect the well to come back to its original specific capacity. There needs to be a determination of the changes in specific capacity due to a lower static level. There is also a question as to whether the reduced production is due simply to a lower static water level or due, at least in part, to losses in permeability in the porous media due to plugging. In other words, when the well is loosing its static water head, rehabilitation may not bring it back to original. If it is due to plugging material, effective removal of the plugging deposits can restore the capacity of the well. There are no absolute calculations that can be made to determine the percentage of lost capacity due to plugging material versus the percentage of lost capacity due to dewatering the aquifer.

There are examples of lost static water level. For example, there are basins in the U.S. where extraction rates far exceed recharge rates. Some of these are so severe that the aquifers have been dewatered by about 800'. This lower static water level has been equivalent to approximately 40 feet of drop a year, and in some cases, is still operated in this manner. This is a practice that can create many additional problems. There are many other parts of the U.S. that were operating without good management practices and have since enacted laws to stop or slow the rate of this happening. Once good management practices are put in place these basins can recover. These items of management are discussed here as they relate to the creation of well problems and the difficulties that are experienced in solving problems.

Some of the problems with extraction rates exceeding recharge rates include:

Exposed perforations: In some situations these exposed perforations can be hundreds of feet. Hundreds of feet of screen are now above the water level and can experience cascading water problems. Cascading water leads to problems much more rapidly than if the screens are below the water level, due to introduction of oxygen into the water and the surrounding formation. One of the parameters that most commonly leads to losses in production capacity and water quality problems is oxygen. Anything that you do to introduce oxygen into a well system will significantly enhance problems in general. If you have cascading (falling) water into a well it becomes very aerated. If you are cycling a well (e.g., starting and stopping a pump very frequently), will also increase the oxygen entering the well. You are essentially increasing your exchange capacity. Every time you have drawdown and then it recovers, you bring a column of air inside the casing to increase your exchange capacity between the water column and the air. Oxygen is the most important parameter in the operation of wells that can lead to problems. During research for my Master's thesis, I found oxygen to be the most significant parameter leading to the biofouling of wells. From an increase in 1 mg/l oxygen the biofouling and the plugging rate were found to increase about 10 times faster.

3

GROUNDWATER MICROBIOLOGY

SOURCES OF BIOFOULING MICROORGANISMS

This will be dealt with at the conceptual level in order to give an understanding of the importance of microbes in groundwater. The first step I would like to take is to dispel the idea that iron-related bacteria infect wells. That is not the case; iron-related bacteria do not infect wells. A well that experiences iron-related bacterial problems does so from natural indigenous bacteria. It is more important that the environment be created for naturally occurring iron oxidizing or related bacteria to proliferate. It is the environmental conditions that are created for the enhancement of these naturally occurring bacteria that will determine if a well experiences an "iron bacterial" problem or not. Many different types of bacteria that could cause the typical depositions of iron (which are commonly recognized as iron bacterial problems or plugging) are abundant in subsurface environments.

There has been a lot of research which supports how widespread microorganisms are in the subsurface environment. Although rare, there may be some situations when one type of microorganism may be introduced by such things as well-drilling equipment. This can prove to be an essential part of the community of microorganisms which could then cause plugging. When added, the missing "ingredient" can cause the plugging to occur. These relationships can be referred to as consortia.

When we put a well in the ground, we basically create an active conduit (i.e., the well) through which oxygen can penetrate down into the ground where oxygen would not normally exist. Now the well starts to be pumped and, in so doing, conditions are created where the flowing water begins to enhance the ability of those naturally occurring microorganisms to grow in and around the well. A zone is therefore created around the well to stimulate these naturally occurring bacteria. The greater volumes of water passing into the environment over the surfaces bringing with it various organic- and inorganic-based nutrients enhance the potential for growth and deposition. The bottom line is that bacteria do not have to be introduced to create these problems.

The U.S. Department of Energy conducted a ground breaking study on the depths and extent to which microbes can be found in the ground and what their nutritional status was. This was of particular interest to the environmental industry in terms of the ability of these microbes to remediate (biodegrade, bioaccumulate) various hazardous materials. These studies help to answer some of the questions posed in the water well industry. Four separate holes were aseptically drilled down close to several thousand feet in South Carolina. Very exhaustive tests were conducted on the samples retrieved from that investigation. All sorts of protective measures were introduced to eliminate the possibility of contamination. Tracers were used to determine if any contamination existed, and if it did the samples were invalidated. It was found was that there were 10^8 to 10^9, which corresponds to one hundred million to one billion bacteria per gram of soil of the samples taken from all depths. That means there are hundreds of millions of bacteria for every gram of soil! Most (approximately 90%) are attached to surfaces in the form of biofilms. When you think about taking a pumped water sample, think about the data, think about the fact that you are taking a pumped water sample, you are taking a water sample, you are not taking a sample from the biofilm where most of the microbes reside within biofilms (Figure 2). We are only measuring the "tip of the iceberg" — perhaps only 10% of the microbes that are actually there!

Subsurface Bacteria

- Indigenous populations typically contain 10^8 to 10^9 bacterial cells per gram
- Approximately 90% are attached and form biofilms

Figure 2

When there is a detachment of bacteria from that biofilm, now there would be some degeneration in the bacteriological water quality and this could generate an unacceptable (unsafe) result including total coliforms. The pumped water sample does not reflect the bacteria in a biofilm. It can create false negatives in that the bacteria are in the well environment but not detected. Anything that causes the detachment of bacteria from the biofilm will therefore cause degeneration in the water quality. What has to be remembered is that a pumped water sample is not representative of the types or numbers of bacteria that are likely to be found attached on a biofilm within the well.

Another fact observed was that depth has no influence (Figure 3). There were just as many bacteria at depths of thousands of feet as there were near to the surface (the water table was approximately 60'). Other investigations in oil wells have shown very high populations at tens of thousands of feet. In other words, depth, as a function, is not the determining factor for bacterial numbers, and depth will not control the population of bacteria that may be present. The key factor for the presence of life is the requirement for water whether that be in a liquid, solid (as ice) or gaseous state. There is literally a flood of micro-organisms since there are not going to be any limitations to growth if there is availability of water, even in such extreme environments as would occur at these depths.

Subsurface Microbiology

- ❑ **Depth has no influence**
- ❑ **Higher numbers are found at the water table**
- ❑ **Highly transmissive zones have 2 to 4 orders of magnitude higher numbers than low permeable zones**
- ❑ **Culturable vs. nonculturable are 4 to 5 orders of magnitude higher numbers**
- ❑ **95% of the isolates are aerobic**
- ❑ **95% of the isolates are chemoorganohetero-trophs**
- ❑ **4,500 different types of bacteria have been isolated from 60 samples in 4 wells**

Figure 3

Figure 4 represents an overview of some of the diversity found in the subsurface environment.

Subsurface Microbiology, Bacterial Groups
- 95% nonstreptomycete
- 3.5% streptomycete
- 1.6% fungi
- 81% rod shaped
- 86% Gram negative
- 67% pseudomonads
- 3% overlap between aquifer mo's and soil mo's

Figure 4

What was found was that the highly transmissive zones were found to have two to four orders of magnitude higher bacteria than the low transmissive zones. Basically water is the carrier for the elements of life which now become dependent on the organic and inorganic nutrients (Figure 5). The greater the ability of water to move through the subsurface environment and the more the nutrient loading, the greater will be the potential to support life (Figure 6).

Influences on Bacterial Abundance Include
- Available Organic Carbon
- Nitrogen
- Phosphorus
- Sulfur
- Moisture
- PH
- Electron Acceptors
- Grazing by Predators
- Immigration of Microorganisms from Other Habitats

Figure 5

> # Geological Influences on Microorganisms Include
>
> - ❑ **Nature of the Geological Stratum**
> - ❑ **Mineral Type**
> - ❑ **Particle Size Distribution**
> - ❑ **Texture**
> - ❑ **Hydraulic Conductivity**

Figure 6

CURRENT BACTERIAL TESTING METHODS

Another problem in groundwater microbiology in general is that the nonculturable bacteria exceed the culturable by a four to five order of magnitude. That gives us a false sense of security for the numbers generated from the test procedures such as the heterotrophic plate where we are determining the numbers of bacteria. The numbers obtained for the culturable bacteria would be much lower than the true number, which would involve the nonculturable.

We have used 500 colony forming units (cfu)/ml as the threshold indicator for the deterioration of water quality. While we are not able to grow most microorganisms from aquifers that have such diverse nutritional requirements and they would not all be able to grow on the surface of a semi-dry agar plate, yet we are expecting them all to grow. Well, they are not all going to grow! We have tremendous differences between those microbes we are able to grow and those we are not able to grow. In recent years, there have been a lot of comparisons done between a heterotrophic plate count and another technique using adenosine triphosphate (ATP) analysis. All of the living microbial cells have a consistent amount of ATP in them and so the amount of ATP in a water sample relates to the numbers of microbes in the water. To do this test, take a water sample and bombard the sample with an enzyme called luciferase; it gives off a glow of light (luciferin) which can be detected by sensitive light detection systems. By the amount of light detected, it is possible to back-calculate the amount of ATP and the numbers of active microbes in the water sample. The limitations of growing many of these bacteria on the plate count are seen when comparing plate count results versus ATP results. For example, with plate count you might have 10 cfu/ml or maybe 1 cfu/ml or 500 cfu/ml, but with the ATP system, the equivalent counts for the same sample may be as high as sixty million, ten million or one million or hundreds of thousands per milliliter. This shows the inadequacies of some of the common testing procedures and clearly demonstrates that ATP would be much more sensitive than the plate count.

Obviously there is a tendency to grossly underestimate the numbers of microorganisms using the plate count techniques, and this gives a very false sense of security. There is clearly such a huge difference from what is counted using an agar plate technique and what is there in reality. This inaccuracy using the traditional techniques, which is measurable in several orders of magnitude (underestimated), is clearly more than many traditionalists would be prepared to accept.

There is a 500 cfu/ml plate count threshold which can be supported as an indicative technique to be used to show the deterioration of water quality from the wells, distribution systems and so on. This indicates that the surfaces in the water system are becoming "dirty." It is really as simple as that. If we start having increases in microbial numbers and there are occasional "unsafe" coliforms or occasional increases in heterotrophic plate counts, it can be a good measure that the system is biofouled and needs to be cleaned. The system I am referring to is any surface within that water environment, whether it be distribution lines, column pipe, pumps, well screens or the surfaces of the gravel pack and aquifers. They need to be cleaned so that there is a reduced risk of detachment of that type of material (e.g., biofilms) from the surfaces. In addition, 95% of the isolates were aerobic (able to grow using oxygen). Again that was done as a part of the Department of Energy study which was a very intensive study by a large team of researchers. This reinforces that it is oxygen that is the most significant material that we add (often inadvertently) to the well. In the absence of oxygen, these bacteria while predominantly aerobic are now able to utilize other electron acceptors such as nitrate, carbon dioxide, sulfate, and oxides of minerals instead of oxygen to be able to "breathe." All of these alternative electron acceptors that the aerobes can use only yield approximately 10% of the respiratory potential (and energy) that oxygen can achieve, and so oxygen is by far the most significant stimulator of microbial growth in wells and aquifers. Once you put a well into the ground and add oxygen into the subsurface environment then you significantly enhance (nine-fold increase) the energy that can become available to the bacteria. Oxygen also often determines the zone of biofouling and/or mineral deposition. The anaerobes, with the exception of the sulfate-reducing bacteria, do not cause the same scale of problems that the aerobes do. Even though the anaerobes cause some problems with corrosion and taste and odor, they tend to form a smaller mass of deposits. The deposits that can be formed by the sulfate-reducing bacteria are iron sulfides.

In the studies to date, commonly 95% of the isolated bacteria are chemoorganoheterotrophs. All that term means is that the micro-organisms utilize organic molecules; it's as simple as that! That is also important because there are a lot of misconceptions about the "iron bacteria." The iron (related) bacteria are still believed by most people to be able to grow on iron. That is not true for the most part. One bacterial genus which has this capability is called *Gallionella*. It has a ribbon-like tail. This makes it very easy to spot under a light microscope and yet, in reality, its importance is really minuscule and it does not cause most of our iron deposition problems. Most of our problems are caused by non-filamentous slime-forming heterotrophic bacteria that do not have to have iron to grow on. The iron they accumulate is more for protective purposes. In other words, you cannot use the iron concentration as the determinant of whether you are going to have iron bacterial problems. The iron concentration will determine the rate at which you will have deposition but doesn't determine if you are going to have iron bacteria.

People tend to focus too much on the filamentous iron-related bacteria that can be relatively easily seen under a light microscope. I have come across example after example where the laboratory reports a negative for iron bacteria based upon light microscopy, and also the observation of the filamentous iron-related bacteria. Yet in the real world it is obvious that there are extensive iron deposition problems primarily driven by high heterotrophic bacterial numbers associated directly with those deposits. Many laboratories are too preoccupied looking under a light microscope to find the filamentous types of iron-related bacteria, while the problems are being caused by chemo-organoheterotrophs when they are utilizing organic molecules. It is also important to understand what causes the rate of (biological) fouling.

At the Savannah River site, 4,500 different types of bacteria were isolated, described and identified from just the four deep wells at that site. These 4500 different types were isolated from 60 samples in 4 wells. There is clearly no shortage of microorganisms! In investigations at the same site working in a so-called pristine (undisturbed) area, investigators found *Enterobacter* and *Citrobacter*. Both of these genera belong to the coliform group of bacteria. One of the first things I will insist upon doing when dealing with insafe (coliform detected) well waters is bacterial identification to genus and preferably species taxonomic level. This is because, from experience on many hundreds of systems, approximately 90% of the time these coliforms are naturally found in the water environments including the subsurface. Of those 4,500 strains, 67% were found to belong to the *Pseudomonas* species. *Pseudomonas* species are the most common microorganisms; these bacteria are ubiquitous, very common slime-forming bacteria, and they are found all over the place! They are very capable of filtering iron out of the groundwater; it's as simple as that. Most of our minerals that are deposited within and around wells, whether from iron, manganese, calcium, magnesium or silicates, are done by biological filtration processes. This is no different to what happens in a trickling filter in a wastewater application. The bacteria are filtering organic nutrients and they are filtering a variety of minerals or inorganic species primarily for protective purposes. Physical protection against such events and the potentially toxic hydrogen peroxide which they naturally produce. Most of the minerals get precipitated in wells as a result of iron-related bacterial activities and are really biological filtration products in the zone around the well that has become enhanced. That is why it occurs for several feet into the surrounding formation because of the availability of oxygen and the access to organic and inorganic nutrients which create on the surfaces an environment that stimulates the bacteria naturally in the formation to start to grow. These growths then start to filter materials in the biofilms formed by the bacteria.

The nature of the geology is going to determine what becomes dissolved in the groundwater. For example, calcium bicarbonates are going to dissolve in the water if the formation contains carbonates. If the water travels through iron- or manganese-rich deposits, some of these minerals will be dissolved into the water by reductive processes (also often bacterial). Essentially the mineral content of the water reflects the history of the groundwater that is moving through the aquifer. This will then determine the water chemistry. Particle size and texture will determine the hydraulic conductivity. This will then basically determine the ability of the subsurface environment to transmit water and therefore support more life. Quite a few years ago I had the opportunity to see a picture of a well that had been literally cut off as a cross section after years of service to see where the fouling was occurring around a well. This was when some dewatering wells around an iron ore strip mine were being removed. Cross sections of the wells were cut out while examining the gravel pack and the surrounding natural formation. The iron-related bacteria, you would think, would be all close to the well in a concentric circle. That is not the case. What actually happens are "finger-like" growths moving out through the gravel pack into the formation. These "fingers" vary in length, one might be only 6", the next one 3', and so on. This is primarily because the water flowing toward the pumping well is not taking even flow paths. It is taking variable flow paths as it moves towards a pumped well. Wherever the flow paths of water are occurring is where there is an enhanced zone for fouling and is where the filtration process is taking place, because water is the carrier for the elements of life. This is one of the difficulties that we face in water well rehabilitation because of the path of least resistance that (treatment) chemicals will take. It is these flow paths which these chemicals are going to go into, that is, the parts of the aquifer that are still open. This is why physical agitation is most important in a treatment process — to overcome some of the limitations caused by the chemicals moving through flow paths and bypassing the plugged zones of fouling.

OTHER ORGANISMS IN GROUNDWATER SYSTEMS

While most emphasis is given to bacteria in plugging and degradation events, these are not the only organisms which can be found in a well. There can also be rotifers, diatoms, nematodes, protozoa, algae and fungi. One of the problems is the way in which you determine whether a well is under the influence of surface water. This has been done in the U.S. extensively over a number of years. If they have surface water infiltration, then they have to be treated as if they were surface water supplies. This means that you have to apply flocculation, sedimentation and filtration. Most people are therefore very concerned to determine whether the well is subject to this surface water infiltration because it means that the water treatment process is going to have to be scaled up to surface water standards.

SURFACE WATER INFILTRATION

A part of the problem faced by the industry is the development of standards for determining whether wells are under the influence of surface water. These can be reasonably used to determine this surface water infiltration factor. I have personally been involved in reviewing results from sampling wells of different depths and diameters. Some of these samples were collected from shallow wells (from less than 50') and some were collected from deeper wells (greater than 800'). The wells less than 50 feet deep are often suspected to be under the influence of surface water. A whole range of organisms beyond the usual total coliforms were determined and it was found that there was no difference! What I have to point out is that there is a very diverse microbial ecosystem in groundwater and wells, mostly naturally occurring. There can be some introduction but most are naturally present. The simple fact that you find protozoa in subsurface environments is not really that surprising since the protozoa are going to be the primary scavengers of the biofilms. Thus the protozoa are recycling the nutrients that were bound up in the biofilm. The nutrients are recycling. Because you have protozoa present does not mean that you are automatically going to have the *Giardia, Cryptosporidium,* and *Entamoeba histolytica* which are the ones that are of concern for health purposes. It means that you have some normal, naturally present protozoa.

4

LONGEVITY OF REHABILITATION TREATMENTS

ACCUMULATION OF MATERIALS IN WELLS AND SYSTEMS

Deposition can become very extensive and is determined primarily by the environment created for biological growth and subsequent mineral oxidation. This deposited material in wells logically affects the production of a well and may also affect water quality. River-connected collector wells tend to have ideal growth environments and can experience extensive buildup of deposits. Certain radial screens in radial wells feeding into a central cistern (caisson) will have more deposited material than others, due to the creation of different environments in the same collector well.

At a collector well in South Dakota, the caisson had a 14' diameter to a depth of approximately 100'. At the bottom are horizontal screens, which project from 200' to 400' out around the caisson like spokes on a wheel. This particular design of a collector well uses a partial river recharge and can lead to extensive deposition. They can plug at a rapid rate due to the creation of an ideal environment. Material not only builds up in the gravel pack and aquifer but also on the well screens, which can very severely restrict flow.

Wherever there is a water environment and there is a surface within that environment, there is always the potential for some kind of deposit to develop on that surface and cause loss of hydraulic capacity and/or water quality problems as the deposition builds up on the surface (Figure 7). These surfaces can include a well screen, the surface of the gravel, or a surrounding formation, column pipes, distribution lines, water treatment equipment, storage tanks, anywhere there is a surface there is the potential for some types of buildup to occur that could then cause problems.

ZONE OF FOULING

The area in and around the well that is being affected by plugging and deposition problems can be referred to as the zone of fouling. This

---

Problem Sites for Biofouling:

- ❑ **Water Wells**

- ❑ **Distribution Lines**

- ❑ **Water Treatment Equipment**

- ❑ **Cooling Towers**

- ❑ **Air Stripping Towers**

Figure 7

usually extends for several feet out from the well. The size of the zone of fouling (deposition) is determined primarily by the availability of oxygen. If you are operating injection wells, have cascading water problems, or if you are starting and stopping your pumps more frequently, this will cause oxygenation to move much further out from the well. This will drive the plugging mechanism much further out into the surrounding formation.

Injection wells can cause particularly severe problems. Injection wells can exist as a number of types. For example, those used for contaminated groundwater extraction go through some type of air stripping system and the water is then injected back into the ground. Those wells can develop cycles of fouling (rehabilitate — pump — clog), which can be as short as one week, so that they become operational nightmares. The plugging is very rapid because of the high concen-tration of oxygen along with high concentrations of organic nutrients as well.

There are also a lot of problems with higher concentrations of oxygen in Aquifer Storage and Recovery (ASR) wells used for water banking. Very many of these systems are being installed in the U.S. and what these are doing is taking water through the winter months into storage, banking (storing) it in the aquifers. Thus when there is a low demand for water in the winter, it is being stored in aquifers. When the demand rises in the summer months, the water is recovered. These aquifer storage and recovery wells will plug at a more rapid rate and will tend to drive the plugging mechanism much more deeply into the formations due to the greater amount of oxygen entering with the water being injected.

Most oxygen coming into a well system does not come from water arriving at the well from points of recharge source, whether that be 500', a mile, 10 miles or 100 miles away. There could be residence times for the water in the aquifer that would be measurable in thousands of years. Most ground waters are devoid of or very low in oxygen. Most of the oxygen entering the well enters down the well column itself. That is not always the case. We do have wells that are river connected and the water is highly oxygenated from the river itself, and they are often used as a means for artificial recharge of the aquifer from the river itself. Since most wells are practically devoid of oxygen and most of the oxygen comes down the well column, oxygen determines the rate and the zone of fouling.

RATE OF IRON DEPOSITION

The concentration of iron in the water is not so important in determining if you are going to have iron deposition. It will, more importantly, determine the rate at which iron deposition will occur. I have on many occasions been involved in the rehabilitation of wells that have a nondetectable iron concentration or a concentration below the secondary MCL of 0.3 mg/l in the water and yet have a lot of iron deposition problems! What is happening in these cases is that it will take a much longer time due to the filtration of larger volumes of water to achieve the same iron buildup when compared to a well with higher iron concentrations entering the well.

There are some wells which will naturally have iron concentrations of 1, 2, 5 or 10 mg/l, and even as high as 75 mg/l of iron in some recovery systems. Where the iron concentrations are that high then there is much more microbial activity leading to the reduction of iron and then to the subsequent possibility of much more iron deposition (oxidation). If you have 5 to 10 mg/l of iron in the water, it does not mean necessarily that there will be many more iron-related bacteria present. It does mean that the buildup of iron deposits will potentially be that much faster and much more difficult to deal with!

RATE OF BIOFOULING

This is important to understand when trying to get an understanding of the potential longevity of a well and longevity of a well rehabilitation treatment (Figure 8). Everyone that practices rehabilitation on specific water wells often finds that the period between treatments will often become shorter and shorter as time goes on. For example, a well may go twenty years before the first rehabilitation, then only maybe ten years before it again requires another treatment — after that treatment, maybe five further years, then three years, and then two years and so on until it is no longer worthwhile to rehabilitate the well. Why does this happen? I used to think that this was because we were not getting satisfactory disinfection of the well to kill the iron bacteria. I no longer believe that! It is not so important to get good disinfection as it is to get rid of the material causing the loss in production. The fundamental goal is to achieve effective deposit removal. I will often call this removing the "CRUD" from the well. The word "CRUD" can mean Completely Removing Underground Deposits.

Rate of Lost Capacity or Biofouling:

□   **Environmental growth conditions determine the rate of biofouling and plugging**

□   **Rate most significantly determined by organics and oxygen**

□   **Biological active zone often takes years to filter enough material from the groundwater**

Figure 8

The rate at which the well looses capacity is largely and most significantly determined by the amount of oxygen and also will be determined by the amount of organic material and the dissolved minerals available (Figure 9). A biologically active zone often takes years to filter enough material from the groundwater to significantly affect capacity and warrant rehabilitation.

Oxygen coming down the well column will determine the zone of deposition, and the reason it takes a long time (e.g., 20 years) for the

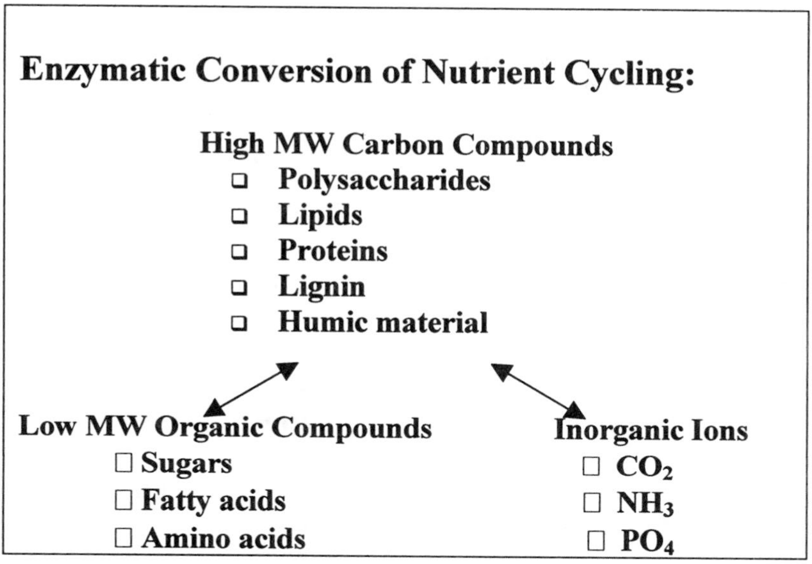

Enzymatic Conversion of Nutrient Cycling:

High MW Carbon Compounds
- ❏ **Polysaccharides**
- ❏ **Lipids**
- ❏ **Proteins**
- ❏ **Lignin**
- ❏ **Humic material**

Low MW Organic Compounds
- ☐ **Sugars**
- ☐ **Fatty acids**
- ☐ **Amino acids**

Inorganic Ions
- ☐ **CO_2**
- ☐ **NH_3**
- ☐ **PO_4**

Figure 9

well to lose its capacity will also be affected by the limitation in available organic molecules. In water supply wells, organic molecules (in other words, food sources for bacteria) are relatively low and it takes a long time (many years of filtration time) for bacteria to filter out enough material to cause growth. Once this material is filtered it may take a long time to cause enough mass of material to cause flow decrease and water quality problems (Figure 10). Significant growth can cause a loss in capacity in that well. Remember that minerals will also be filtered out with the organic material to add to the significant mass of deposited material. Mineral content present in a deposit is, most of the time, the larger percentage of the overall mass of material, normally consisting of approximately 75 to 80% of the material.

Wells with very high organic loads, which can cause rapid growths and losses of capacity, are the groundwater recovery wells. These wells are used to extract organic contaminants such as hydrocarbons or solvents. From experience, the loss in capacity (rate of deposition) on those wells may be a hundred-fold more rapid, when compared to water supply wells. Wells can plug in a week's time compared to twenty years' time for a water supply well because there is no limitation on organic molecules as the primary feedstock. These organic contaminants are ideal growth nutrients for the microorganisms; the microbes do not have to filter for twenty years and can now grow very quickly. Very extensive sliming and biofouling problems can be seen to occur on these types of wells, which now often become operational nightmares.

Longevity of Successful Treatment, factors

❑ **During rehabilitation, 100% of the deposited mass is often not removed**

❑ **Bacteria can regrow very quickly on organic material left behind after treatment**

❑ **Key to increasing time between treatments is: EFFECTIVE DEPOSIT REMOVAL**

Figure 10

LONGEVITY OF REHABILITATION TREATMENTS

During rehabilitation it is difficult to remove 100% of that deposited mass which is in the well (Figures 10 and 11). If you do not remove all of that mass of deposit, then the microbes do not have to filter for such a long time. For example, if only 50% of the material was removed, then 50% remains resident in the well. That means regrowth would occur much more quickly the second time around and what happened in years may now be repeated in months! The key therefore to increasing the time frame between treatments gets down to effective deposit removal. Anything that can be done to efficiently remove all of those deposits will increase the length of time before the next rehabilitation will have to be done.

In my opinion, even more important than that is the problem that many wells may have excess production capacity. Some impedance of the water flow by deposited material may not be to such an extent that it will influence flow, since the well is not at its production capacity. Looking at porous media, for example, sand or gravel, we have a lot of void spaces and there may be partial plugging of some of these void spaces before it starts impacting the capacity of the well because there is excess capacity.

Longevity of Treatment, concerns

- ❑ **Many wells have excess production capacity**

- ❑ **Flow of water through an aquifer may not be impeded by some deposition of material**

- ❑ **If all deposited material is not removed, it does not take as long for plugging to fill the void spaces in the porous media**

Figure 11

That may also be exaggerated by excess capacity that comes from the depth of the screened areas or producing zones due to variations in well hydraulics. There can be some zones in the well which start getting plugged while some of the other producing zones now begin to make up the production being lost from the plugged zones of the well. It is conceivable to have a lot of deposition in the well and formation before there is an impact on the specific capacity. There comes a critical point when you no longer have that excess capacity and you now start to have turbulent flow losses; there now develops an identifiable loss in the specific capacity. Some of the pore volume can become plugged before flow of water changes out of laminar flow. If that well is now rehabilitated, you may get that specific capacity back to 100% of original but not remove 100% of the deposited mass that is in the well. If you have not completely (100%) removed that mass, it does not take as long for the well to begin to plug up again because all of the excess capacity does not exist. Once the well is back in production, material is again being laid down over the surviving plugging material and the well's capacity will decrease again more quickly than if all of that mass had been removed. There is no longer that surplus capacity (due to the mass not removed) and so the loss in capacity will occur that much quicker. Another way to visualize this is to consider a pipe with some deposition of material not impacting the flow of water through the pipe. Once enough deposition exists then flow will be impeded, but if all the material is not removed, it will not take as long to deposit additional material and impact flow. It is important to get to the original surfaces when cleaning wells and water systems. The bottom line here is: EFFECTIVE DEPOSIT REMOVAL.

WATER WELL REHABILITATION BENEFITS

One of the basic benefits is restoring lost well specific capacity (Figure 12). The common comment is that "I can no longer get enough water from my well any more," and that means that you are not only unable to pump as many gallons per minute but the cost of pumping water is higher because of high electrical costs (Figure 13). Electrical cost of pumping water is the single most significant cost of water. Municipalities and industries are finally beginning to look more and more at well efficiencies as one aspect in producing the water. This involves not just looking at capital expenditure for wells but also how many dollars it is costing to pump a specific amount of water.

Well Rehabilitation Benefits:

- ❑ **Restoring lost capacity**
- ❑ **Decreasing pumping costs**
- ❑ **Extending life of the well**
- ❑ **Solving water quality problems**
- ❑ **"Safe" bacterial samples**

Figure 12

You can decrease your pumping costs by keeping the pumping water level much higher. If you are pumping at a certain number of gallons per minute and you suddenly begin to have plugging of the well, the drawdown will become bigger as the pwl declines. When you experience a bigger drawdown, the pump is now pumping against more head. If the pump was not selected to pump against additional head and it moves off the most efficient point of the curve, it costs more to operate. Therefore it is costing more dollars to pump that same amount of water out of the well. This can become very significant and it can also be used to economically justify water well rehabilitation.

Figure 13

Sometimes the costs of well rehabilitation sounds expensive but you have to look at how significant and how short the period would be to get the payback period. Here is an example of a city in California. This city has a well that is approximately 1,100 feet deep, 16" diameter, with the costs of the rehabilitation about $60,000. That sounds like a lot of money to spend on a well, but the payback (cost recovery) for that $60,000 only took eight months, just from the savings in electrical pumping costs alone! This was due to the prevention of as much draw-down during pumping. Most often the increased water production is taken into account but the savings in power consumption are often not.

It is possible to look at rehabilitation of a well just to keep the efficiency up and the pumping costs down and nothing else. It is now very simple to run cost-benefit analyses to justify the rehabilitation just on the basis of the payback you can expect! It becomes a "piggy-back" effect, and if we were to show that it would take fifty years to recover the costs, then clearly, do not do it. It cannot be justified by that means; there are other benefits which also, obviously, have to be taken into consideration before the decision is made. If it is the electrical costs that are driving the need to rehabilitate the well, then the decision is very simple in terms of acceptable payback times.

The loss of efficiency of a well can involve plugging.   In another example where, before a rehabilitation treatment, the well was pumping at 1,200 gpm with a drawdown of 81' to give a specific capacity of 14.8 gpm/ft, the power costs for every million gallons of water was $104.59. The capacity after was 2,580 gpm with the same drawdown of 81' but it is now possible to pump much more water. The specific capacity is now 31.9 g/ft and the costs of power per million gallons dropped to $70.47. When this energy saving is taken into account, the results from rehabilitation in power consumption alone amounted to $12,300 for the first year, so savings can be very significant.

What is frustrating is that most people do not even look at the costs of pumping water most of the time. That is now starting to change with higher cost of electricity. Rehabilitating a well more frequently during the early stages of plugging can be justified through preventing the loss of that well.

You can prevent the well from becoming so massively plugged that it becomes almost impossible to ever restore the capacity again. Routine rehabilitation of that well not only removes the plugging but also allows the well to remain operable for a longer period of time. Those plugging deposits which can plug the well up become so extensive, so massive, and so hardened and cemented that it can become almost impossible to ever get those deposits out of there!

ECONOMIC ANALYSIS OF WELL REHABILITATION

One common question when a customer is faced with the cost of a rehabilitation is "wouldn't it be cheaper to drill another well?" This question is asked again and again. This question, however, always has to be addressed. You have to do an economic analysis to compare the costs of rehabilitation with the costs of replacement. Most of the time, rehabilitation remains the most attractive because the cost often represents 10 to 30% of the replacement costs on wells. You also have to consider complete replacement costs which includes moving pumping equipment, new well housing, electrical connections, and relocating lines. In some parts of the U.S., to get a permit issued to drill, it is now common to have to undertake a wellhead protection or source protection study in order to ensure that the well is not going to be impacted by contamination or some other possible compromise. There can be a number of additional costs, which might not be recognized immediately when determining the cost of putting in a new well.

Rehabilitation most often becomes a very attractive economic option when all of these additional factors are considered. Even though you do not have a 100% guarantee of success, it has to be based upon experience and confidence that the rehabilitation process selected has a good possibility of working. If there is not a high confidence that the rehabilitation will be successful, then there is a natural reaction of the customer to move towards the replacement option. This would offer a higher confidence that an adequate supply of water could be achieved and it would have to be done eventually anyway.

A cost-benefit analysis also has to be done where it is probable that the new well will require rehabilitation treatments over the years. A cost-benefit analysis can help to determine this. There are also regional variations in the important factors that come to play in the decision. It should be remembered that if the wells have to be rehabilitated every six months, then how much longer would a new well last before it, too, would require rehabilitation (perhaps two years), and so there are going to be major regional differences which will affect the cost-benefit analysis. For example, in the midwest states of Indiana, Illinois and Kentucky, some wells require treatment every six months to every year because of iron-related bacterial plugging and mineral deposition problems. This means that costs could be $5,000 to $15,000 a year to keep those wells going, but not even the new wells will last long before they start to suffer from the plugging and deposition problems. You are not going to get away from those problems because you are not infecting the wells with iron bacterial problems, because they are naturally present, and you may simply be buying time by drilling a new well. There are things that can be done to control the loss of capacity that will be discussed later. That is all a part of cost-benefit analysis.

5

IDENTIFICATION OF WELL PROBLEMS

BENEFITS OF PROBLEM IDENTIFICATION

In addition to well construction and aquifer characteristics, it is essential to determine the nature of the plugging material. Determining the nature of the deposited material allows better selection of rehabilitation chemicals and/or procedure. Well construction and aquifer characteristics relate more to the development or redevelopment procedures on a well. It is most important to understand the well construction, such as screen type, gravel pack, etc., and the type of development before the appropriate redevelopment method can be selected.

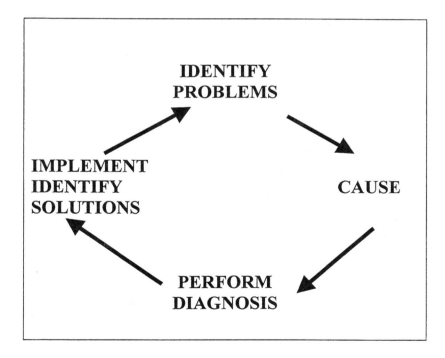

Figure 14

The approach that we most often take to improve well rehabilitation is to properly identify the problems related to each well rather than simply lumping them all together (Figure 14). It is better to look at each well separately. Figure 15 shows some of the problems that will be discussed. To start identifying specific problems on a well it is important to look at any historical records (e.g., specific capacity, water chemistry etc.) that may exist and evaluate the operational aspects so that you can properly identify what the problems are and how significant they are. It may be surprising, but one problem in the water well industry is that we may come up with one treatment specifically for one well and it is automatically adopted around the world simply because it was effective upon one well!

IDENTIFY PROBLEMS

- **Lost Capacity**
- **Turbidity**
- **Corrosion**
- **Red Water**
- **Taste / Odor**
- **Pump Wear**
- **Water Quality Fluctuations**

Figure 15

There can be tremendous variations in plugging deposits, as will be described later, since many different types of mineral encrustation and content of plugging deposits will justify the need to define what the problems are. This all means that it is important to clearly identify what the problems are so that you can properly select what you can do on individual wells. It does not mean that you will have to do something different on every well because similarities in plugging deposits exist geographically and geologically. There are going to be regional similarities and there are going to be regional differences. The similarities will be based on water chemistry, upon bacteriology, upon the types of formations and well construction. This means that you are not going to have to have customized well treatment for every well in a well field because there will be similarities within those wells.

LOST CAPACITY

The average length of time within the U.S. that a well will naturally loose some of the capacity and requires a first rehabilitation treatment is about twelve years. Some wells will loose capacity in less than a year. Some wells will go fifty or sixty years without loss in capacity! This means that the average life of a well is about twelve years before a rehabilitation treatment is needed. Also once treated, why do the time intervals between subsequent treatments become shorter and shorter and shorter? And surprise, in my opinion, it is not due to the fact that we are not getting good disinfection of wells! It is because of our inability to remove deposits. You can obviously have plugging deposits that can exist in the gravel pack and for some distance into the surrounding formations (Figure 16). Most of the plugging deposits that we have seen extend for several feet away from the well screen or the face of the open hole into the surrounding formations (such as the gravel pack and aquifer). We do not commonly have such extensive plugging deposits further out than that, which is fortunate for it is increasingly difficult to remove deposits that far out! We do have certain rehabilitation processes such as Aqua Freed^{TM} that have much better penetration capabilities and can overcome rehabilitation limitations of other rehabilitation procedures.

There are some techniques better able to penetrate deeper into the formations than other techniques, and these will be described later. These techniques can overcome to some extent the limitations imposed by having the plugging extend deeper into the formations.

TURBIDITY PROBLEMS

Turbidity problems can be caused by a variety of things in water. The common causes of turbidity include, air entrainment, sand, suspended solids, and high bacterial loads (Figure 17). Some of these issues will be discussed in more detail later.

Problem: Lost Capacity

Figure 16

CORROSION PROBLEMS

Again, I approach corrosion problems in the same manner as any other problem on a well or water system. The cause of corrosion can be a variety of things, and so the first action is to identify the cause of the source of corrosion specifically. It is necessary to identify the source of corrosion in order to select the correct solution. Is it in the materials used for construction, or something in the water that is corrosive by itself, erosion corrosion, electrolysis, etc.? Some of the corrective actions taken to control corrosion include impressed current or cathodic protection of the coating materials on the pipe and choosing materials less corrosive. The common types of corrosion seen in wells are electrolysis, electrochemical corrosion or microbially induced corrosion (Figure 18).

Problem: Turbidity

Figure 17

Electrolysis can be fairly significant. This type of corrosion is essentially when your well or pump is used as a grounding mechanism.

If you have a high voltage electrical field from an over-head power line or a transformer, for example, you can have an electrical field around that well or pump. This can actually be measured, and they are called "stray currents." These are utilizing the well or the pump as a grounding mechanism.

Figure 18

There have been times when we have been able to separate a discharge pipe from a pump and measure an electrical potential across those two when they are separated. This indicates that the well is becoming a grounding mechanism. You can have fairly rapid losses of metal which are fairly easy to solve. You can use dielectric couplings or isolation flanges that will electrically isolate both the pump and the well and so prevent that electrolytic corrosion from happening.

Most corrosion is started due to the formation of an electrochemical cell. All that means is that you have dissimilar metals; one of those metals becomes an anode (anodic) while the other metal becomes a cathode (cathodic) and these now establish an electrochemical cell. There is now the transfer of electricity from the anode to the cathode. You can now have the dissolving of one of those (anode) and the build up on the other (cathode). That activity is also going to be dependent upon the amount of total dissolved solids (TDS) in the water. The TDS will determine the rate of transfer of the electricity in general. TDS may be a better determining parameter for corrosion than the saturation indexes. There are many other things in water that can also create the electrochemical cell. For example, the presence of hydrogen sulfide is corrosive because it is often anodic in relation to the cathodic metal surface. We can therefore establish the electrochemical cell just by having the presence of hydrogen sulfide. A deposit existing on one part of a surface and no deposit on another part of the same surface can also create a differential potential and lead to corrosion.

Microbial induced corrosion (MIC) will account for about 30% of our corrosion problems. Most predominant in this are the sulfate-reducing bacteria (SRBs). Most of the time that is evident through the formation of some kind of deposited material. Most of the MIC is commonly referred to as "under deposit corrosion" because the deposit is necessary for the creation of the anaerobic environment. The SRBs are anaerobic bacteria. These SRBs can grow in the anaerobic environments commonly created by the iron-related bacteria and the slime-forming bacteria. They are growing underneath that deposit (such as a tubercle or nodule) within biofilms. The aerobic bacteria in the growths are growing on the surface and are creating an anaerobic environment underneath where the SRBs can flourish.

When a tubercle is examined, it will often be seen to have a reddish-brown hardened shell with a black inner core. Beneath that black slime when that is scraped off, you will see pitting of the pipe underneath. That pitting has been the result of the establishment of microbial corrosion and it is the SRBs that are actually causing the corrosion by forming sulfuric acid and hydrogen sulfide. Pitting is often the most destructive form of corrosion because you can have failure of a system with only a little loss of material. This little loss of material is concentrated at a "pit". The pitting can also become autocatalytic. Once a pit is established there is a differential potential between the bottom of the pit and the surface. For this reason the pitting can continue after the other corrosive agents have been removed.

Microorganisms can also grow upon surfaces and create differential potentials across surfaces. This can now lead to corrosion. This will also create an electrical potential and initiate corrosion. There are also some microorganisms that are able to concentrate chloride ions. They are called chloride concentration cells that will then be a corrosion agent with the action based upon the concentrated chloride.

To combat microbial corrosion you have to prevent the deposits from forming. Whether you have to use biocides, chlorination or whatever, to prevent the formation of the biofilm, will then prevent the deposits from developing and the anaerobic corrosive micro-environment. We do not have nearly as much anaerobic corrosion taking place in anaerobic environments as we do in aerobic environments. The aerobic environment is actually a better environment for the anaerobes because they prefer to grow underneath those tubercles. This becomes a better environment than the strictly anaerobic. This gets down fundamentally to the form of microbial interactions and the fact that these can be working in a synergistic (or symbiotic) relationship — in other words, they are helping each other out. The SRBs are not very capable of utilizing large organic molecules, they are relying on the iron-related and the slime-forming bacteria to break the larger organic molecules down so that they can feed on the products of that degradation — the smaller simpler organic compounds.

RED WATER PROBLEMS

Often, when a well is pumped, there is an initial discharge of red water (Figure 19). That red water is simply caused by the detachment of material due to the high velocity water moving through the formation and perforations. After a time such as ten minutes, it clears up due to flushing the detached material with no subsequent detachment of deposits. Many of the "unsafe" bacterial samples, including the total coliform to atypical to TNTC, may be due to the fact we are having occasional detachment of that normally 90% attached material.

Red water is a relatively common problem spearheaded by complaints about colored water, stained laundry and toilets. It is very important when you are talking about red water problems to identify which part of the water system is actually causing it. If you are just going to rely on customer complaints to determine whether you have a red water problem, and you automatically assume that it is the well, you may not get to the site of the problem and you may not get the problem under control with well rehabilitation. You may have the establishment of tuberculation inside the distribution lines.

Problem: Red Water

Figure 19

Remember that anything that is going to lead to the detachment of that material may cause a red water problem (Figure 20). You can also have that coming from a well where the initial discharge is discolored to a red or brown for the first minute, or five minutes or ten and then it clears up. This is primarily due to the detachment of material into the water during that initial high velocity state.

The act of shutting the pump down and then starting the pump up again will create a high velocity state. This will lead to the detachment of that normally attached material. That event indicates that the well needs to be cleaned — it is as simple as that! The material formed by tuberculation, slimes and growths needs to be removed so that when you create the high velocity conditions, there will not be any detachment of that material.

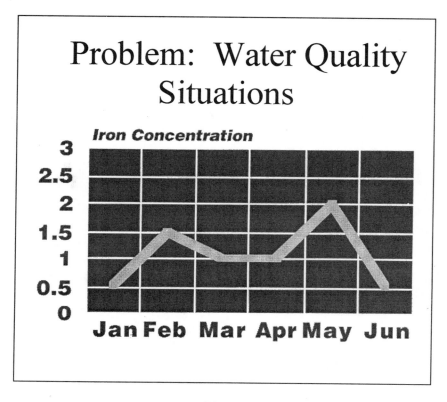

Figure 20

If we take a look at tuberculation, or nodule formation, it is a 100% biological process. Having said that, though, this material has a higher mineral than biological content (normally 75 to 80%). Throwing chlorine at the nodule is not going to remove it because most of the structure, although biologically created, is composed of minerals. Understanding what it is composed of is more important when trying to remove it than determining whether it is biological or not. Understanding the mechanisms of plugging is more important if you are trying to prevent the material from depositing. From the hundreds of deposit analyses the most common percentage in those deposits that is mineral scale is 75 to 80%. About 20 to 25% of those deposits are going to be either organic or organic fractions. Therefore, while it is biologically filtered material, the bulk of the material that we will be removing will be mineral in content. Consequently, we have to lean more towards better acids or formulated chemistry to dissolve the mineral content rather than targeting something like shock chlorination to disperse a little bit of the bacterial slime. Chlorine is not going to remove materials. It is as simple as that!

You can also have increases in the iron content, which is actually dissolved iron (ferrous-Fe^{++}). You can have changes in water quality as the well gets older because as you operate these wells over the years you are going to be filtering dissolved minerals out of that groundwater.

When you build up enough minerals in this oxidized zone, there will be a lot of iron and manganese becoming trapped around that well. Once you have enough of that oxidized iron trapped around the well you can then start having iron-reducing bacteria which will then dominate in a reduced environment. They will then take that oxidized (ferric-Fe^{+++}) iron and reduce it back into the water to the clear dissolved (ferrous) state. The iron-oxidizing bacteria first took the dissolved ferrous (Fe^{++}) and oxidized it (aerobically) to the insoluble ferric (Fe^{+++}) forms, which entered the mineral deposits. The iron-reducing bacteria then reduced the insoluble ferric (Fe^{+++}) and anaerobically converted it to the dissolved ferrous (Fe^{++}) state. Thus, after a number of years of operating a well, you can suddenly get increased iron content in the water as the dissolved ferrous (Fe^{++}) forms begin to enter the water again. These will oxidize in the presence of oxygen to again make red water. We have done rehabilitation on wells which have been successful in removing mineral deposits and plugging only to get the dissolved ferrous (Fe^{++}) in the water under control and back to ambient concentrations. Some of these will again reappear after a period of time to reduce the water quality. After one treatment, the concentration of iron will actually increase due to additional detachment of material and exposure of the reduced environments. This challenge is very difficult to handle and it has sometimes taken several treatments of removal before we have been able to bring this type of problem under control, back to the original iron concentration.

SINGLE WELL WATER QUALITY VARIATIONS

Another reason that you can have significant changes in water quality from a single well, and this is something of greater significance than we have previously recognized, is you can start having a different blend of water coming from the same (single source) well. The reason for this is that the water entering through various parts of the screened regions may blend differently as some parts of the screen plug up. For the example provided from the spinner survey (in the well hydraulics section), most of the water from the well is produced through the upper section of screen. This would have one particular quality of water and then this screened section would begin to plug. Water would now be drawn from a lower section of screen and thus be drawing water from a different aquifer, which may have very different characteristics. As the upper screen plugged, then so the produced water quality would change closer to that of the water coming from the lower aquifer. Thus there would be a different blend of water from the same well affecting the water quality. This concept is demonstrated by observing the spinner survey results before and after well rehabilitation. The quality of water from the well is going to be determined by the blend from the various waters mixing from the different screens down that well which are each being subjected to a different rate of plugging. This will be more obvious if you are screened against multiple aquifers using a screen—blank—screen—blank—screen approach then the differences can become very significant. If a well is completed only in one aquifer with a single screen passing down through the aquifer, it is less likely to have significant water quality changes. A well may still experience sand pumping problems or "unsafe" bacterial problems as the blend of water changes in a single screened aquifer due to higher velocities in zones that may have historically had zero to low velocity conditions.

Color problems in water can sometimes arise just because of such an event. As some screens or parts of a screen in the upper part of the well become plugged, the well begins to make up the water from the deeper formations. If now one of these deeper formations has a colored water problem (such as tannins and lignins) then this would now be reflected in the water quality coming from that well. It is just the different blending of water in the well that can sometimes, and dramatically, cause a color problem in the water. After a well such as this one with a color problem has successfully undergone rehabilitation, the blending of the water now returns to the original pattern and the colored water problem disappears. The solution therefore is effective rehabilitation, because once you get the removal of the plugging material, blocking off flows from water that is not colored, then the blend will change back to its original acceptable characteristics without any color.

ODOR PROBLEMS

There is a range of odor problems that can occur in water wells (Figure 21). Common odors are the "rotten egg," earthy–musty, fishy smells, petroleum odors and pond smells (Figure 22). Most of the odor problems from wells are due to some kind of biological growth. The most common odor which many people are familiar with is the "rotten egg" type odors caused by hydrogen sulfide gas generated by the SRBs and is very common in groundwater systems. Again, you cannot simply rely upon shock chlorination to control these types of problems because the SRBs are growing very much under the protection of the mineral contents of the tubercles. To kill these SRBs sheltering under the tubercles, you must use effective means to dissolve the tubercles and expose the SRBs and then you can get a much better disinfection. If dealing with the "rotten egg" / hydrogen sulfide gas problems, the simple application of shock chlorination may not control the problem.

Figure 21

Taste and Odor Problems

- ☐ **"Rotten egg" odor**
- ☐ **Earthy–musty**
- ☐ **Pond smell**
- ☐ **Fishy smell**
- ☐ **Petroleum odors**
- ☐ **Septic odors**

Figure 22

Earthy–musty (geosmin and isoborneols) odors do occur in groundwater systems. It is not the blue-green algae (cyanobacteria) which cause the problems as they do in surface waters; it is a group of filamentous bacteria called the Actinomycetes or the Streptomycetes that can produce very typical earthy–musty smells by producing the same chemicals as the cyanobacteria. They establish themselves within biofilms, within distribution systems and within the ground waters. They are a little more difficult to kill; I have dealt with a few situations but they are not that common. The problem is that they produce spores, which are more difficult to kill, so that a second disinfectant treatment is needed routinely. The second treatment has to follow the first when the surviving spores have germinated (after the first treatment) and the cells are now vulnerable. To further demonstrate this point we can look at how the actinomycetes are grown in the laboratory. We will heavily chlorinate a water sample to kill all the vegetative cells and the other microorganisms and then allow to spores to germinate and produce the actinomycetes.

Petroleum odors when detected automatically require that a hydrocarbon scan be done to determine whether the odor is coming from petroleum based products. If it is not (no hydrocarbons detected), the cause may be some bacteria which can produce similar odors to petroleum. The most common bacteria to produce these odors are the pseudomonad bacteria. They can produce very typical petroleum-type odors and, again, we have dealt with those over the years.

6

UNSAFE BACTERIAL RESULTS

SOURCES OF UNSAFE SAMPLES

When laboratory results come back showing that the water is unsafe due to high total coliforms, I have found in approximately 90% of the cases that infiltration of coliforms from the surface is not the cause – 90% of the time the total coliform positives are not due to true contamination of the well water from the surface.

Unsafe (bacteriological) water quality issues are becoming some of the most critical issues being faced in the U.S. at this time. The issue of "unsafe" bacterial samples due to coliforms and high bacterial counts has become a more critical issue in recent years, partially due to changes in regulations. The frequency of wells experiencing "unsafe" bacterial samples and the concern of this issue is increasing. A most common scenario is that an "unsafe" result is returned from the laboratory for a given sample and a repeat sample is tested which then comes back negative (absent). This is often put down to a sampling error on the first sample and the water is now considered safe. Most of the time it is not sampling error; most of the time it is because of detachment of normally attached biofilms resulting in variations in the natural loading of the pumped well water with these bacteria.

The reason that the bacteria could be present at one point in time and absent at another is because of attachment and detachment. Understanding the attachment of bacteria to surfaces and the occasional detachment is paramount to understanding many of the water quality problems experienced in water systems. Both analyses were correct; the first was positive and the second was negative. Most people forget to keep in mind that bacteria growing within water systems are attached to surfaces. The reason why there is an increasing occurrence of positives in samples is that the bacteria are detaching, coming into the water sample and being detected in the sample with increasingly sensitive testing procedures.

When there is occasional detachment, resampling will often not detect these bacteria because they are no longer detaching from the surfaces! Most of the time it is not sampling error; most of the time the samples are simply reflecting whether the bacteria are attached (absent from the sample) or detaching (present in the sample). It is most important to understand this when taking samples from a well since this is a pumped water sample and is not totally representative of the populations attached to surfaces in both quantity and bacterial types. Sometimes solving these problems can develop into absolute nightmares as we try to achieve "safe" samples with total coliform negative results. This means that it is necessary to work on wells often for months before we are able to maintain a "safe" sample schedule. In my opinion, the single most important aspect in understanding and solving these problems involves hydraulic limitations.

These hydraulic limitations result from the inability to pump the stagnant zones or low flow zones which are most commonly at the bottom part of the well. The bottom part of a well can become a "no-flow" or "dead" zone in a well and this is the biggest source for wells having "unsafe" samples (Figure 23). In distribution systems, such as we have in wells with detachment of tuberculation, detachment of biofilms can also result in the occasional "unsafe" sample.

Other Sources of Bacterial Problems

- **Pump setting in a deep well**

- **Attached vs. detached bacteria (sessile vs. planktonic)**

Figure 23

TOTAL COLIFORMS

There is now a much more heightened awareness of the problem with total coliforms. It does cause a lot of municipalities a lot of problems involving more applications of disinfection or more rehabilitation treatments. As far as "bad" samples are concerned, they are called bad samples because the water has failed the total coliform or the fecal coliform test. Fecal coliforms are particularly of concern because this means that there is probably *E. coli* present and there is probably a direct contamination taking place. This could be coming from livestock, a septic tank or septic system, improperly sealed well or some kind of direct contamination that is taking place. It is always indicative of direct contamination when you are dealing with *E. coli* or *fecal coliforms*. In the majority of wells that fail a bacterial test, it is rare to find the presence of *fecal coliforms* or *E. coli*. *E. coli* would, in fact, be a better indicator of actual fecal contamination.

The total coliform test has been used since the turn of the century (1910, I believe) and it is still used fairly extensively today as an indicator of contamination (Figure 24). But I no longer believe, of course, from the experiences that I have had that it is an indicator of contamination. I tend to believe that most of the total coliform problems that we deal with are due to natural indigenous bacteria. Even though the bacteria are naturally indigenous, it does not mean that we are going to have all of the wells with these types of total coliforms. It does not mean that we are going to have them present all of the time.

Total Coliforms

Definition:
Comprises all aerobic and facultatively anaerobic, Gram negative, non-spore forming, rod-shaped bacteria that ferment lactose to acid with gas within 48 hours at 35°C

- **Includes fecal coliforms**
- **Are usually nonpathogenic**
- **Presence of coliforms indicates potential presence of pathogens associated with waterborne disease outbreaks**

Figure 24

There are variations in the microflora present in water environments. These bacteria are all a part of the enhanced natural processes that can lead to lost capacity as well. When we are operating new wells, we may not have any growths detaching. However, this will change when the wells become biologically enhanced and growths begin. There are going to be some of the total coliforms such as the *Enterobacter*, and the *Citrobacter* (the ones we most commonly deal with) are also very capable of iron filtration and deposition as a part of the iron bacterial biofouling. They are going to be a part of the iron-related bacterial group; they are not going to be a separate group.

Some of the studies done by the various research organizations talk about the "harboring" of the coliforms in the distribution systems, "harboring" of the coliforms being basically the protection of the coliforms within the biofilms. I am not convinced that they are "protected" by or being harbored in the biofilms. I suspect that they are a part of the biofilm formation; they are equally responsible for the buildup of materials (such as the tuberculation and mineral deposits) as much as the other microorganisms active in the biofilm, and then they are released by detachment. They do not need to be protected – for they are a part of the biofilm. Both the total and the fecal coliform tests are routinely used to determine whether a water sample is safe or not.

We also often have to deal with the TNTC, overgrown and atypical results. TNTC means "Too Numerous To Count." All that means is that when a total coliform test is done using a membrane filter, there are too many colonies growing to be able to count properly and so it is considered that there are too numerous colonies to be able to count accurately. This results in a failure.

We also deal with "overgrown" or "atypical." In these cases, there is interference by a multitude of different bacteria that often results in the ability to read the results (Figure 25).

Bad Bacterial Samples

- **Total coliforms**
- **Fecal coliforms**
- **TNTC** (too numerous to count)
- **Overgrown**
- **Atypical**

Figure 25

Because there are so many heterotrophs growing and interfering, we cannot trust the test data that we get because of all of the foreign interference. Occasional TNTC, occasional overgrowths or atypically, the occasional total coliform can primarily be due to the result of the detachment of normally attached microorganisms from the deposits and tubercles, which are now getting into the water sample. Whenever dealing with an "unsafe" bacterial problem we must consider the possibility of surface water infiltration, and we must always rule out that possibility prior to rehabilitation.

Once we have established that there is a bacterial problem, it is always important to proceed on to specific determination. We need to precisely determine which type or types of bacteria we are dealing with. Are we dealing with one that should not be so widespread in groundwater systems? Are we dealing with enteric bacteria such as the *Enterobacter*, the *Klebsiella* and the *Citrobacter* that are much more common in groundwater systems? Sampling techniques, earthquakes, biofilms, and natural indigenous bacteria – what are the critical factors to explain the data (Figure 26)?

Occasional or Persistent Bacterial Problems Sources:

- **Surface Water Infiltration**
- **Sampling Techniques**
- **Earthquake**
- **Biofilm**
- **Natural Indigenous Bacteria:**
  - *Enterobacter* **sp.**
  - *Citrobacter* **sp.**
  - *Klebsiella* **sp.**
  - *Aeromonas hydrophila*

Figure 26

When you do a pump repair on a well, you could pull a pump and replace it with a sterile pump. Most of the time, you are going to get an "unsafe" sample for total coliforms after you do a pump repair, not because they were introduced with the pump (it could even have been sterile) but because you have disturbed the natural attached bacterial growths down the well. You can actually have a higher number of total coliforms and heterotrophic bacteria after chlorination because you are detaching some of the normally attached microorganisms because of the activities involved in the chlorination! Instead of having the (normally) 10% detached microorganisms in the system, the total may go up to 50% detached and we are pumping a lot more in the pumped water sample. It is primarily dependent on the detachment of the normally attached microorganisms. *Aeromonas hydrophila* accounts for roughly 30% of the positive samples on total coliform analysis.

ANALYTICAL PROCEDURES

Another reason for the increased frequency of occurrence in the numbers of positive total coliform tests is the adoption of new criteria and testing procedures. For example, in the presence/absence testing, we now have a zero tolerance, whereas previously there was an allowance (e.g., <5% of the samples positive) before a well was considered to have failed. We also have the defined substrate media like the Colilert™ and the Colisure™ that are actually more sensitive in their ability to detect these microorganisms that were often missed in the past with some of the membrane filters and the most probable number techniques (Figure 27). We no longer miss these; we do have more sensitive techniques. I have made a general observation that when the Colilert™ test is used, an increase in the number of "unsafe" bacterial samples from industrial and municipal systems occurs. Split samples from some wells have been done, with one sample going to a laboratory that is using the traditional membrane filter or a MPN method and the other going to a laboratory using a defined substrate medium such as the Colilert™. The Colilert™ test will often indicate a failure (unsafe) while the traditional technique will pass the water as safe. This does support the possible increased sensitivity of the test. That is some of the cause for our increase in the numbers of unsafe samples.

PROBLEMS IN ACHIEVING SAFE SAMPLES

If fecal coliforms are absent from water sampled but there are total coliform bacteria present, I would suggest that these should be identified to species. If you do have a positive fecal coliform or *E. coli* then you are dealing with a true contamination problem. A "magical" treatment of the well is not going to cure that problem until you go to the source of that contamination and eliminate it. Whether that source be septic tank or feedlots, you have to determine that source before you can effectively treat the well. If the problem is total coliform, go to the identification and determine which species are present. Most of the time, we end up dealing with *Enterobacter cloacae, Citrobacter freundii,* and *Klebsiella oxytoca* as the most common species in the total coliform positives. There are others such as *Serratia* but these are the ones most commonly found. These are indicators that they are naturally indigenous. Also you need to eliminate the potential contamination sources − look at the overall area. Examine also the well construction. Does it have grout seal (annular seal)? Does it have a good well seal? Are there any surface bodies of water around or does the formation contain a confining layer? Is the well shallow or is it deep? Is there a potential for holes in the upper casing to allow surface water or perched water to enter into that well? All of these things need to be addressed or eliminated.

Coliform Analytical Methods

Based on the presence or absence test routines, the following tests are approved for total coliform analysis:

- ❑ **Multiple-tube Fermentation Technique**
- ❑ **The Membrane Filter Technique**
- ❑ **The Presence-Absence (P-A) Coliform Test**
- ❑ **The Colilert^{TM} System (MMO-MUG) Test**

Figure 27

The next step is to find out the history in terms of what has already been done, what strengths of chlorine treatments have been used, whether there has been success at all, what other chemicals have been used. The bottom line here is to determine what the limitations have been to getting good controls and disinfection efficiency of the coliform problems in the past. Once this has been done, then strategies can be developed to improve the disinfection efficiency and get the problem under control. Strategies that can be used include overall better disinfection, greater volume, greater concentrations, better choice of disinfectant, better pH and whatever other strategies are being applied. If you fail to improve the situation, then the next step is to take the same approach as you would for a well that has lost specific capacity. You are probably going to have to deal with a problem that requires deposit removal since that is where the total coliforms are living. Chlorination has clearly by now failed to achieve the elimination of the total coliforms and so the blended chemistries of acids, surfactants and disinfectants need to be applied at the level normally used for the more effective removal of deposits or in a preventative maintenance treatment. If the disinfection of the well doesn't work, then the pump will have to be pulled and a much more aggressive treatment applied. I have been involved in many projects where this had to be done to ensure the removal of the deposits harboring the coliform bacteria. What was often found was the well could be cleaned of the problem by including adequate flushing of the bottom (static or low flow) parts of the well.

HYGIENE RISK SURVEYS ON WELLS

The Centers for Disease Control and Prevention (CDC) undertook an investigation of approximately 5,600 private wells in the midwest states after the flood of 1993. The reason for this investigation was the need to examine the health risks associated with wells, and the flood provided the focus and the financing (from the federal emergency funds). A ten-mile grid of these nine states was used as a sampling protocol, where one well out of every ten-mile grid was sampled. As a result, this survey was not limited by such factors as where there were rivers. What was found was that approximately 43% of these wells were total coliform positive. Some of these wells also were found to be *E. coli* positive. Some of these wells were simply hand-dug brick-lined in-ground cisterns, and it is not surprising that these types of wells could have problems. When we take these wells out of the survey and just use the properly constructed bored sealed wells, approximately 31% were still tested positive for total coliforms.

Several other states have also performed surveys, where approximately 25 to 56% were total coliform positive. At the National Groundwater Association meeting in Las Vegas in 1996 there were two and a half days of sessions on the biological aspects of groundwater. There was at this meeting discussion from several states that did some of their own investigation. They included wells that were deep or shallow, wells that were large diameter (example 18") and wells that were small diameter. They also had wells that were sand point wells that were literally driven into the ground for some 20'. These sandpoint wells have a 2" diameter wire wrapped screened well with a point on the end. Those were found to have the lowest total percentage of total coliform positive problems! This had everyone at the conference scratching their heads for the answer as to why the sandpoints would have much lower total coliform problems? I believe that it fundamentally comes down to the effective flushing of the well. The sandpoint well should not have any stagnant zones. These wells should not have any stagnant water and every time the water is extracted, it is effectively being flushed.

The wells that were found to have the highest percentage of total coliform problems were large diameter wells, because when you are pumping 5 gpm of water out of a well that has a large diameter screen and casing there might be 5,000 gallons of water in the well itself. That water may sit there for weeks or months. That is predominantly a stagnant zone. That is were the bacteria can become active. One of the first things that I will look at to determine the cause of these "unsafe" samples is well hydraulics. What is the possibility of stagnant or low flow zones in the bottom of the well? How many times has a video-inspection of a well shown that. When you're looking at the videotape of a well, very often you will see that the well screen zone is very clear at the upper part of the well. As you descend down the well it becomes cloudy, then turbid and murky and when you get to the bottom of the well, you can't see anything because it is so cloudy and murky. This is primarily because there is no flow, as the bottom part of the well is possibly stagnant and there is not enough upward velocity. You can get some occasional "unsafe" samples and odor problems when the water comes from these stagnant zones. These are more likely to occur when there is a partial plugging of the producing zones. Here, the lower zones now start to making up some of that lost production, and then we start having enough up-hole velocity in lower parts of the well to start lifting some of the stagnant material up from that static zone in the bottom part of the well. It is no different than a dead end in a distribution system.

This is a fairly significant problem in the water well industry. I am part of a task force on the bacterial contamination in water wells, supported by the National Ground Water Association (NGWA). This is a part of the investigation that was done by the CDC. The CDC concern is primarily health significance in order to determine whether these occurrences of unsafe total coliforms had a true, health-risk significance. The CDC wants to determine the significance of the biofouling of wells from the point of view of the potential risks to human health, particularly in view of the approximately 31% of wells showing total coliform positives. It was found that there was no increase in gastro-intestinal illness and other related diseases. One of the directions that this task force on bacterial contamination of water wells was taking was they wanted to start a massive sterilization process on wells.

As a microbiologist on this task force, I did not see how this could be successful. We are not going to be able to totally control the bacteria in the wells. However, what we can do is eliminate the true contamination events. We have to strengthen the well codes so that we have properly constructed wells and improve the preventative mainten-ance on wells to prevent biofilm from forming, and that is what is going to determine success and prevent the occurrence of occasional "unsafe" samples. We have to prevent the biofilm from forming so that there will not be detachment, and there will not be the bacterial releases causing "unsafe" samples. Other than that we may not be able to get these wells under control long term. Yes, we can treat wells that have an "unsafe" sample, and we can get them under control but it is not going to eliminate the problem — just control it. You have to remember that these total coliform bacteria are going to be there attached, and while they are attached they are not going to be detected in the water being tested. They will only be detected when they detach. A negative for total coliforms really means that there are no total coliforms in the water being pumped from the well; it does not mean that they are not down there! We are essentially only measuring the tip of the iceberg with a pumped water sample.

7

WELL HYDRAULICS

FLOW DYNAMICS OF WELLS

This section on well hydraulics is based upon observations of problems and solutions on many wells during the past fifteen years. The concept is being investigated, and more research is needed to more fully understand the dynamics of flow of water in a well. One part of well hydraulics, which is sometimes not appreciated, is that you can have a variation in the production through a well screen even over a relatively short distance (example: 10 feet of screen). Groundwater does not flow evenly into a well along the entire length of a well screen (Figures 28, 29, and 30). There is a variable production with depth even with a short well screen. If you are not pumping very much water, you are not going to have much up-hole velocity. You can also have variations as to where the water is being produced, even in a 20', 15' or 10' well screen. It doesn't have to be hundreds of feet of well screens like we do have in many situations. You do not have to have a long well screen to get variable production with depth.

Constraints: Well Hydraulics

❑ **The bottom parts of many wells are stagnant zones. This is similar to the "dead ends" in distribution systems**

❑ **The lack of adequate "up-hole" velocity leads to water quality problems including "unsafe" bacterial samples**

❑ **The pump needs to be pulled for proper well rehabilitation**

Figure 28

Factors: Well Hydraulics

❑ **Production profiles change as wells age**
❑ **Water quality changes can result from changes in the production profile**
❑ **Well hydraulics influence the effectiveness of well rehabilitation**
❑ **Excess production capacity exists in many wells**

Figure 29

Impacts on Production: Well Hydraulics

Variable production with depth results in high velocity conditions in some parts of the well which can result in:

❑ **Sand pumping**
❑ **Excessive entrance velocities**
❑ **Enhanced biofouling**
❑ **More mineral encrustations**
❑ **"Unsafe" bacterial samples**

Figure 30

We have support for that in the fact that we have installed a lot of suction flow control devices to prevent sand pumping. Sand pumping problems can be due to high approach and entrance velocity conditions or poorly gravel packed wells as water moves toward a pumped well.

These high velocity conditions may exist even though wells are properly designed. You can still have high velocity conditions in some regions of your wells and no velocity conditions in other parts of the same well. Now if we can equalize the flow with a suction flow control device and extended pump suction, we will have more equal flow along the length of the well screen. We have been very successful at achieving this with many different screen lengths. We have been able to stop many sand pumping problems, not because we are physically filtering sand with this suction flow control device but we are changing the dynamics of flow and the hydraulic characteristics so that we do not have high velocity conditions.

Wells are often designed with an entrance velocity of less than 0.1 feet per second. What that means is that the water enters through the screen at less than one tenth of a foot per second. This is a fairly conservative estimate. It is designed below that entrance velocity so you should not have sand migrating into the well because there is not the velocity to carry the sand. That sounds very good in theory but in practice we are calculating entrance velocity based upon well discharge capacity in gallons per minute (Q) and we are assuming that every square inch of the well screen is producing the same amount of water, but it doesn't happen! We commonly have some parts of our well screens that are producing 90% of the water while other parts of the well screen are not producing any of that water. So you are going to have some parts of the well screen that are going to be in a high velocity condition while other parts of the well screen are going to be in a no-flow condition. So even though we can project a theoretical result, we do not necessarily achieve that.

There are things that we can do to change the flow dynamics and flow hydraulics. In fact, we have been very successful in preventing the recurrence of "unsafe" bacterial samples by putting suction control flow devices into wells. This means that every time the pump comes on, it effectively "flushes" the whole length of the screen because the bottom part of the well is no longer the stagnant zone. The extended pump suction eliminates those dead zones and the whole length of the well becomes active. We have been able to also clearly demonstrate increases in specific capacity in certain instances.

Clearly variable production with depth due to different velocities down the length of the well can result in excess sand pumping, excessive entrance velocities, enhanced biofouling and mineral encrustation, and also "unsafe" bacterial samples. The bottom parts of many wells are essentially dead zones similar to dead ends in water distribution systems. The lack of up-hole velocity leads to water quality problems and limitations in rehabilitation. The bottom line is, if you want to get the best cleaning of a well, you need to pull the pump. In many situations we can still get good cleaning with the pumps in place, but you have to understand what the limitations are. It is like taking a five-gallon pail and trying to clean it on the inside with a little brush by just swirling the brush in the top of the pail. We are not agitating it, as I believe one of the most essential aspects of good well rehabilitation is proper agitation. Yes, the chemistries are important, but chemistry does not bring everything into solution. If we were able to take a chemical which had the ability to put everything into solution, then it would be fine, just pump everything off after the treatment. We cannot do that — we have to rely on the agitation and the redevelopment methods. These are the more important aspects of well rehabilitation in general.

Production profile changes can cause changes in the water quality, and excess production can exist in many wells. You can see the shifting in the zones where water is being produced in a well before and after rehabilitation. This can be done using a spinner survey. A spinner survey means that a flow meter measuring velocity is lowered down a well in the entire length of the producing zone, while a well is being pumped. The velocity and rate of flow are recorded along the entire length of the producing zone so that the velocity of water in the well can be determined. This has proved to be a very useful tool and many wells have been monitored this way. This helps to explain many of the features involved in water well hydraulics we have already been talking about. In the example being used here, the 18" diameter well is screened from 260' to 1030'. This is larger than many of the wells you may be familiar with, and while this well has hundreds of feet of screen you can apply this information to a well that has only 20' of screen.

If only one aquifer is screened, then it is unlikely to get water quality changes due to the change in flow dynamics. Here is an example of a well that we rehabilitated in southern California using Aqua Freed™. In this example we have an E log, different productive zones, temperature log (differences can be seen with the temperature with depth), fluid resistivity and the spinner survey results. What I want to focus on in this example are the up-hole velocities and the gallons per minute pumped from each zone. This well was being pumped at 3,000 gpm and, before rehabilitation, the zones were producing (in order of top five producing zones going down the well, gpm) (Figures 31 and 32): 600, 568, 793, 481, and 314.   Let us now look at the same well after rehabilitation with the well still being

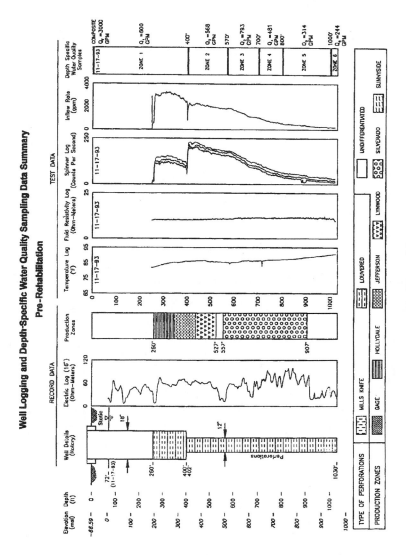

Figure 31

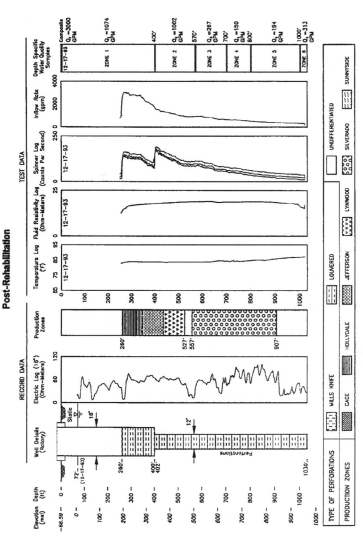

Figure 32

pumped at 3,000 gpm (percentage change from before rehab. given in brackets): 1,074 (+79%), 1,002 (+76%), 267 (−64%), 150 (−69%) and 194 (−38%). The spinner survey after rehabilitation is more comparable to when the well was new. After treatment, the upper zones of the aquifer were now able, once again, to meet the demand being placed on the well and so the lower zones were not contributing as much water. This was presumably not because they were not capable, but because more of the flow was being produced from the upper sections.

As you can see, the post-rehabilitated well is more comparable to when the well was new. As the well again begins to plug then it can be expected that the upper producing zones are going to plug first and again, as in the past, the water now begins to be extracted from the deeper zones. Now it becomes more understandable why water quality changes can come about as the process of plugging and redistribution of water from the different depths of the well occur. We also begin to see support for the excess capacity that exists in a well and aquifer. For this reason early deposition of material may not impact specific capacity.

There are occasions, particularly with the deeper wells, where it has become very difficult to ensure ongoing production without the loss in water quality associated with the shifting in the dominant production zones. One answer has been to attempt to equalize the flow to reduce the high velocity flows from those sections of the screen where sand is a potential problem. We have had more problems in maintaining safe bacterial samples in regions of the country where there are a lot of deep wells because of the more significant stagnant zones and because of this shifting in the peak production zones as the biofouling progresses. There is a need on these wells to create a flushing condition where water is moving throughout the well and there are no stagnant (dead) zones. In one deep well, we had a problem with "unsafe" bacterial results which was resolved when we took out the permanent pump and placed a temporary test pump at the base of the well. For fifteen days, the bacterial samples came back "safe." As soon as we put the permanent well pump back in the upper casing, then the "unsafe" bacterial samples started again. The pump set at the bottom of the well had a flushing action to prevent the formation of stagnant zones and, more importantly, flushing the detached microorganisms from the well and surrounding formation. The answer is to keep flushing actions going in the well to prevent dead stagnant zones forming to support bacterial growth. It was not until we installed approximately 700' of extended pump suction to the bottom of the well that we were able to control these conditions. This way, every time the pump comes on you effectively flush that well.

I do consider understanding well hydraulics one of the most significant aspects of well rehabilitation — understanding why we have to pull pumps in order to clean wells, because we need to get access to allow good agitation and flushing of the detached debris. We need to get very good up-hole velocities to lift all of the deposited material that is gathered in the bottom part of that well and the surrounding formation. Remember that many times you cannot just shock chlorinate a well to take care of an "unsafe" bacterial problem. You will not be able to remove the mineral scale. Disinfection is an important part of treatment, but remember that there is too much protection for the microorganisms within that mineral scale to allow success just by disinfection most of the time. The chemistry necessary to cause the dissolution of that mineral scale has to be an important part of the treatment before you can get to, and disinfect, the bacteria. Yes, shock chlorination may give you a safe bacterial sample immediately after treatment, but that does not mean that the problem is solved.

POSITION OF BIOFOULING RELATIVE TO WELL SCREEN

When video logs are taken down-hole, there may be a tendency for the water and screens around the very active producing zone to be cleaner. As the production is being lost from a particular zone (such as the upper zones in the example I just discussed) then the screens will become visibly plugged and the biofouling and mineral deposits are extending back into the pack and formation beyond the well screen. At the bottom of the well where there is a possibility of more stagnant zones, the water will appear murky, cloudy or even black and yet, at this site, the biofouling and mineral deposits may not extend as far back into the formation. The screen and pack here is not producing and so there would be very little enhancement of the environment to increase biological activity and mineral deposition. The camera being obscured by all the material in the well water does not mean that the gravel pack is heavily biofouled; it means simply that the water is stagnant water with a poor water quality.

The concept is therefore that where there is a producing zone, it can be expected that the biofouling and mineral deposition will extend further back beyond the gravel pack and into the formation. And yet, that may not be so evident on the well screen itself until the latter phases of the fouling. In a nonproducing zone, particularly a stagnant one, then much of the biological and chemical activity will take place right in the well and not extend far into the formation beyond the well screen. You have to be able to correctly interpret what you see when you look at the video of down-hole camera inspections. All you can really see is inside the well itself, and the visual observation may be misleading but still an essential tool in well problem identification.

SUCTION FLOW CONTROL DEVICES

These are installed in a number of different ways. Basically there is a casing with a well screen or open hole below the casing. The pump is most often set at some point in the upper casing of the well. Because limitations in hydraulics are imposed in the bottom part of the well and there are often sand pumping problems, suction flow control devices can be considered. The first thing that we do when there is a sand pumping problem is to determine whether there is a hole in the screen or casing with a video inspection. Once you have eliminated that possibility, then you have to determine whether the problems are due to high velocity conditions. Some of those increases in velocity may be due to the fact that you have partial plugging. This partial plugging could mean that you now have water being produced from formations that have never produced water before. You might now be using zones that were never properly developed before. It is even possible to start pumping drilling mud many years after wells have been in service. There have been examples after rehabilitating wells that are twenty years old where we have pumped large quantities of drilling mud from those 20-year-old wells. This would indicate that they were never developed properly and there never had been the velocities necessary to get the fines from the surrounding formation and gravel pack. You can have sand pumping problems due to partial plugging, and then the remaining open areas of the screen begins to get turbulent or high velocity flow. If you have these conditions, then there is the potential to get sand migration into the well. The answer to that situation is rehabilitation to open the plugged screens and get the production back to the original patterns and reduce the entrance velocities.

However, the sand pumping may not be caused by these problems described above. For example, the sand is coming from right at the top of the screen. Consideration should be given to the installation of suction flow control devices. These basically have extended pump suction that have a little open area at the top, but the amount of open area increases as you go down the suction flow control device. As an example, this device was fitted to a well where this was done in

Missouri. The well originally produced 750 gpm, but over the years it began to produce sand. It produced sand as it declined to 200 gpm. We installed one of these suction flow control devices and were able to get the well back up to 700 gpm sand-free with a 57% increase in specific capacity. This increase was achieved because zones of that well screen were now being utilized that had never produced before. That was a 35' long well screen! This illustrates that you do not have to have a very long well screen to get variable production with depth and have sand-pumping problems. To have an increase in specific capacity is extremely significant since it showed that the suction flow control device could significantly affect the way the well could be operated in a positive manner.

Related to well hydraulics, in Louisiana, there are many wells which have screen openings that are only 0.010" to 0.014" of an inch. Some of these wells lose production very rapidly and are so packed that by the time they are five years old, the specific capacity is down to 30% of the original and we have not been able to get them back again. Under these conditions it is time to start taking the approach of over-designing wells. This means both in terms of diameter and production rate. For example, a well which could pump at 3,500 gpm may be pumped at, for example, 1,500 gpm with, more importantly, extended pump suction, suction flow control devices or flow equalization devices in place. These suction flow control devices would be used to reduce the entrance velocities impacting the surrounding formations that carry the fines that lead to the mechanical blockage problems. Now, the advanced recognition of these potential problems, such as mechanical blockage can be taken into account and the well design and operations modified by the application of these types of devices.

8

MICROBIAL TRANSFORMATION IN OXIDATION AND REDUCTION

NITRATE PROBLEMS IN WELLS

Nitrates are the number one problem causing the shutdown of water wells (Figure 33). The number one cause for taking wells off-line is the presence of unacceptable nitrate levels. Again, this is a regional problem. For example, it is more common in Nebraska and Kansas where we have shallow aquifers and oxidized environments (aerated) in agricultural areas. Under these conditions, ammonium coming from agricultural, sanitary, industrial or fertilizer operations will become oxidized to nitrate by the nitrifying bacteria (nitrification). What is taking place here is that the most reduced form of nitrogen (ammonium) is often put into the environment, where it approaches the groundwater environment which, when it becomes oxidative, will be transformed biologically to nitrate, the most oxidized form. This means that there is an eight-electron transfer to make this happen (Figure 34).

Nitrate (NO_3)

- **MCL at 10 mg/L**
- **NO_3 is readily leached though soils**
- **Many different sources exist**
- **Can act as an electron acceptor in place of oxygen for many bacteria**

Figure 33

Nitrogen Cycle

- **Eight-electron shuttle between**
  Most oxidized   Most reduced
  $+5$ (NO_3) ◄———————► -3 (NH_3)
- **Ammonification — no change in valence, NH_3 is released from biomass**
- **Nitrification — ionized NH_3 (NH_4^+) utilized by bacteria under oxic conditions by chemolitho-trophic nitrifying bacteria to NO_3**

Figure 34

The nitrate, when the conditions become anaerobic (reductive) again, will now reverse back toward the ammonium state (ammonification), or the nitrate may be reduced down to nitrogen gas (denitrification) (Figure 35).

Fixation and Release: Nitrogen Cycle

- **Nitrogen Fixation**
- **Assimilation of gaseous N_2 which is reduced to NH_4**
- **Denitrification**
- **NO_3 is reduced to intermediates (NO_2) and lost primarily as gaseous nitrogen (N_2)**

Figure 35

All of these processes are controlled by various groups of bacteria and so we see the nitrification/denitrification cycling to move the nitrogen through a number of forms (Figure 35). Nitrate is generated aerobically when there is oxygen present. A "shuffle" is seen between the nitrate and the ammonium, with the nitrates dominating in the aerobic conditions such as would be present in many shallow wells. Many elements are constantly being transformed both directions between oxidative and reductive environments (Figure 34).

Nitrates can be a very difficult problem and there are some communities in the U.S. which have to face this problem. All of the wells are producing water with greater than the MCL of nitrate and so it is not possible to blend the waters to get the nitrate value into the acceptable range (i.e., <10 ppm). These communities have to deal with treating the water to remove the nitrate, a very expensive process. There are some new in situ systems in which organic sources are added to cause the nitrates to be removed by denitrification; these are still being evaluated. The nitrate is essentially being "breathed" anaerobically by the denitrifying bacteria and broken down primarily to nitrogen gas, which is lost to the atmosphere (Figures 36 and 37).

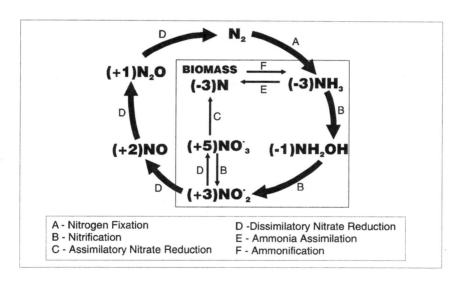

Figure 36

Nitrates can be literally degraded in the ground water. The reason that we do not see nitrates rapidly degraded by denitrification naturally in the ground water is that there is not enough organic source, or the redox may be too high, to allow the nitrate to be degraded.

Denitrifying Bacteria

❑ **Many microorganisms reduce N oxides: (NO_3 to NO_2 to NO to N_2O)**
❑ **When reduction proceeds to gaseous N_2 products such as N_2 and N_2O, it is referred to as complete denitrification**

Figure 37

The symptom causing "blue babies" is high nitrate concentrations in the water. It is actually not the nitrate itself that is harmful to the babies. When we drink nitrate in water, the gut microorganisms of adults and children are able to break the nitrate down to nitrite. Nitrite can also be produced from the oxidation of ammonia (Figure 38). This nitrite wants to become oxidized again and it takes oxygen from the blood (Figure 39). In babies, there is not enough capacity in the blood to satisfy this demand and hence the symptom of "blue babies" is caused because the blood in the babies becomes starved of oxygen.

SULFUR CYCLE IN GROUNDWATER

Like the nitrate-ammonium cycle, the sulfide-sulfate cycle involves an eight-electron transfer (Figure 40). Sulfates commonly occur in water and sulfides are often present, associated with "rotten" egg odors, black water/slime/deposit-problems and corrosion. The sulfur is constantly being shuffled back and forth between the oxidative and the reductive states. The availability of oxygen in the water (redox potential) will determine primarily in which direction the "shuffling" will go. Low or no oxygen would shunt towards the sulfide and high oxygen towards the sulfate. Sulfur oxidation can also take place by an organism *Beggiatoa*, where yellow elemental sulfur globules can occur in large quantities.

Nitrifying Bacteria

Stages in the Nitrification Process:

- NH_4 (ammonium) oxidized to NO_2 (nitrite)   *(Nitrosomonas* sp.)*

- NO_2 (nitrite) oxidized to NO_3 (nitrate) *(Nitrobacter* sp.)

Figure 38

Nitrites (NO_2)

- **Oxides produced by both denitrification and nitrification**
- **NO_2 is toxic to humans**
- **Causes methaemoglobemia ("Blue Baby" syndrome) when NO2 enters the blood and reacts with hemoglobin**
- **Facultative anaerobes like *E. coli* reduce NO_3 to NO_2**
- **More toxic to infants**
- **Can be product from NO_3 in drinking water**

Figure 39

The Sulfur Cycle

❑ **Involves an eight-electron shuttle:**
 Most oxidized   Most reduced
 $+ 6\ (SO_4)$ ←——————————→ $-2\ (S_2)$

❑ **SO_4 acts as an electron acceptor**

❑ **H_2S is produced under anoxic conditions**

Figure 40

The transformation between oxidative and reductive environments is actually no different for iron. In fact, when I was working in Regina, we had a 30-gallon aquarium in the laboratory, which was unlike any other aquarium. It was filled with a gravel pack, there were PVC injection points for injecting air and feeding nutrients, There were vertical screens at either end and water could be recirculated through the pack to get iron bacterial growth. Today we would call that a megacosm. Very quickly the gravel pack in the megacosm filled up with and completely plugged with a brown iron bacterial slime because we were feeding it as well as aerating. These growths could be easily seen through the walls. For several years after it was set up and no longer operated, it proved to be very interesting to observe because the deposits went from a black to a brown and back to black roughly every month in a flip-flop manner. This cycling (brown to black to brown to black) continued for a number of years. Thus, what we were observing there was a flip-flop between an oxidative state (dominated by the red-brown ferric states) and the reductive states which were dominated by the black sulfides. As long as water is available for this to take place, it would most likely last from now to eternity with the recycling of organic and inorganic nutrients.

It is important to understand these types of dynamic events when you are looking at changes in water quality. These very dynamic events may be associated with water quality changes. You need to understand not just the mechanisms that might be involved in each chemical change but also the interactions occurring between these changes under different conditions. Those indigenous microorganisms that are almost inevitably present play a major, if not dominant, role in the shuffling of these elements within groundwater.

9

CAUSES OF WELL PLUGGING PROBLEMS

CATEGORIES

Most of the time, plugging problems in wells can be separated into three different categories: physical or mechanical type blockages, mineral encrustation, and biological type plugging (Figure 41). Although they are separated into three separate categories, there is a combination of these things happening together. Often the bacteria are the "glue" or the "cement" that will filter the minerals from the water and will also act as the "glue" or "cement" that will hold the fines from the surrounding formations as they are moving towards the pumping well. If you did not have something to stick them to the surfaces then the fine sand or silts will often continue to move towards the well and be pumped with the water. Thus, physical blockage and mechanical blockage can result as the fine material moves toward the pumping well and this can become packed up against the gravel pack, the void spaces of the formation, and to the well screen (Figure 42).

IDENTIFY CAUSE

? **Physical**

? **Mineral Encrustation**

? **Biological**

Figure 41

MECHANICAL PLUGGING

It is important to know what type of aquifer the well is in and what type of potential is possible for mechanical blockage because we cannot dissolve that material chemically. We have to rely on traditional redevelopment techniques such as swabbing, mechanical agitation and airlift pumping with high velocity water, and most of the time, we tend to underestimate the degree of mechanical blockage existing. I have found that a lot of the time, people will do a bacteriological test, and, if they find bacteria, then that becomes the entire focus of the treatment while ignoring any of the other mechanisms of plugging. In some ways this is misleading us, by leading us away from the true causes to focus just on that one aspect.

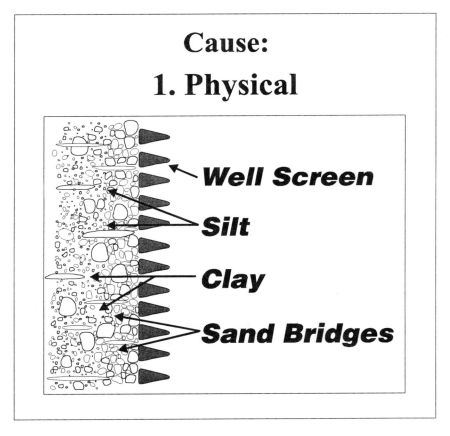

Figure 42

We need to focus on the multidisciplinary aspects of the entire well system and aquifer. I find that people sometimes put too much emphasis on bacterial numbers — that people become much too concerned as to whether there were 13 cfu/ml, or 30 cfu/ml, or 100 cfu/ml! It is not such a big deal. We also do not know if a sudden increase in bacterial numbers is due to a random detachment. It is important how significant these numbers are; the real question is how active are they? We need to get away from numbers and consider ranges of bacteria and how active these bacteria are likely to be and then to look at the "big picture" — of the mineral encrustation and the physical blockage at the same time.

MINERAL ENCRUSTATION IN WELL PLUGGING

The majority of minerals that become deposited in wells and water systems are a result of biological activity (Figure 43). We do not have the environment, most of the time, created to have purely chemical oxidation of iron and manganese. We do not have high enough oxygen concentrations, i.e., high enough pH values to allow this to happen (like the iron from the ferrous to the ferric state).

Figure 43

There is an active metabolic process in which the bacteria are deliberately oxidizing or deliberately filtering the iron from the water for long-term storage to protect against starvation, from peroxide toxicity and against predation from other microbes. The simple fact that shock chlorination has been used for preventative maintenance of wells over the years is support for the fact that mineral deposition is biologically driven. This is because shock chlorination does not change the water chemistry in the aquifer: it's not dissolving the minerals that have already precipitated, the only thing you are doing is preventing the biofilm from forming and therefore you are preventing the biological filtering process from taking place. The water being pumped into the well still contains the dissolved minerals because biological filtration is not happening due to the transient impact of the shock chlorination.

We do not have purely chemical oxidation taking place most of the time. If you have cascading water or more highly oxidized water, yes, you can have purely chemical oxidation. Often the products of that oxidation may not "stick" (become attached — form a part of the plug) and be pumped from the well. Again, the bacteria are responsible for holding these oxidation by-products together within a slime to form the plugging condition. And so it may be seen that it is the bacteria that become mostly responsible for holding these materials (e.g., deposited minerals) and forming the binding plug.

The most common mineral encrustation we find in wells is iron (Figure 44). Much of the time iron forms red to brown types of deposits. We also have a lot of black deposits that are also iron species as well. If we use the example of a tubercle, that nodule from a pipe, that black deposit underneath will often be a reduced form of iron as iron sulfides. There are a lot of people who, upon seeing black deposits automatically think that this is manganese because manganese can form black-type precipitates, but a lot of black iron deposits are automatically diagnosed wrongly as manganese because of the color.

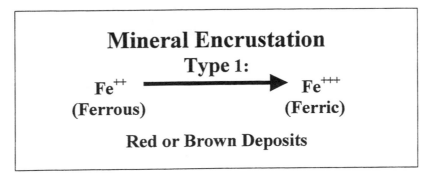

Figure 44

Manganese deposits, like iron, are mostly biologically precipitated (Figures 45 and 46). Manganese requires seven times more oxygen to become oxidized and manganese is a more difficult mineral to dissolve chemically. When exploring ways to remove this mineral content from wells, manganese deposits always present a greater challenge than iron. Manganese problems are common in parts of North America.

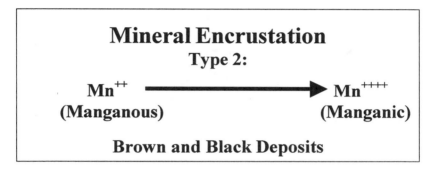

Figure 45

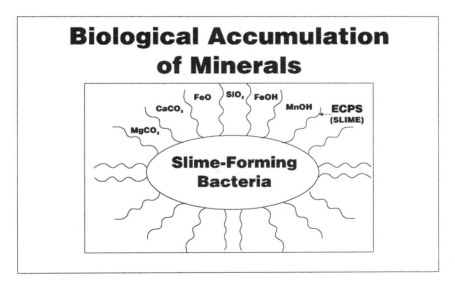

Figure 46

There are also white to yellow type deposits resulting from more of a chemical type of reaction, although you can have some of these reactions controlled biologically. If there are high calcium or magnesium bicarbonates dissolved in the ground water, there is a potential to start the precipitation of calcium and/or magnesium carbonate. The amount of these bicarbonates that the water will carry is very dependent on both temperature and pH.

The pH is predominantly determined by carbonic acid (i.e., CO_2 dissolved in water). This means that the amount of calcium (and magnesium) which is going to be brought into solution is going to be dependent on pH and temperature. When you pump water into a well that is connected to an aquifer that is under pressure, you can start to drive off some of the (dissolved) carbon dioxide by releasing pressure from the formation. This causes the pH to shift upwards into a range where the calcium now deposits as calcium carbonate (Figure 47).

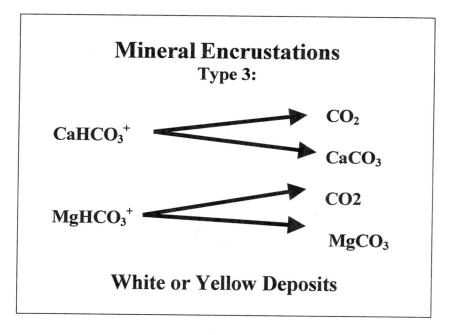

Mineral Encrustations
Type 3:

$CaHCO_3^+$ → CO_2
$CaHCO_3^+$ → $CaCO_3$

$MgHCO_3^+$ → $CO2$
$MgHCO_3^+$ → $MgCO_3$

White or Yellow Deposits

Figure 47

These deposits are no different than the types of deposits that are seen in boilers and in heat exchangers where essentially the same things are happening with the pH changing and the temperatures rising. Some calcium and magnesium carbonate deposition can also occur biologically because there is another group of microorganisms called the chemolithoautotrophs that can actually utilize carbon dioxide as their sole carbon source — they can take the carbon dioxide right out of the water.

Calcium carbonate — the same material that builds up in a kettle — is very easy to dissolve chemically. If all of our deposit problems were calcium carbonate based, then we could almost guarantee 100% removal because we can bring them into solution. It is much easier to dissolve chemically. The more difficult deposits to dissolve and remove are the iron-, silicate-, and manganese-based ones along with some others, particularly when they are combined with sulfates and phosphates.

To look at the risk of mineral deposition, it is useful to look at solubility charts. In ground water, there are a number of different cations (positively charged) in solution, a number of different anions (negatively charged) in solution. Any one of these ions can form into insoluble precipitates (Figure 48). The more difficult deposits to dissolve are phosphate and sulfate species (these often form into denser deposits that are much more difficult to dissolve).

There can be tremendous variations in deposit structures. It is not practical to consider all deposit structures to be the same; there is a lot of variation. In reality, these complexes can be many different percentages and variations of the many different things that are commonly dissolved in water.

Again, there is an automatic tendency to simplify these things too much when we are looking at only one particular aspect of the problem. Some of these complex precipitates can be very difficult to dissolve chemically. There are different acids and other chemicals used in the water well industry to treat problems. These will be addressed later.

TESTS FOR BACTERIA-CAUSING PLUGGING PROBLEMS

About 80% of the problems in water wells are biologically caused (Figure 49). The iron related bacteria (IRBs) are a bacterial group founded upon a number of misconceptions. There are still a lot of people who focus on the idea that all of the IRBs require iron in order to be able to grow, in other words, they get their energy from the oxidation of iron from the ferrous to the ferric state.

Simplified Solubility Chart

Anion → ↓Cation	F-	Cl-	Br-	I-	HCO₃-	OH-	NO₃-	CO₃⁻²	SO₄⁻²	S⁻²	CrO₄⁻²	PO₄⁻³
Na⁺	S	S	S	S	S	S	S	S	S	S	S	S
K⁺	S	S	S	S	S	S	S	S	S	S	S	S
NH₄⁺	S	S	S	S	S	S	S	S	S	S	S	S
H⁺	S	S	S	S	CO₂	H₂O	S	CO₂	S	H₂S	S	S
Ca⁺²	—	S	S	S	SS	VSS	S	VSS	VSS	X	S	—
Mg⁺²	VSS	S	S	S	S	VSS	S	VSS	S	X	S	—
Ba⁺²	VSS	S	S	S	VSS	S	S	VSS	—	X	—	—
Sr⁺²	VSS	S	S	S	VSS	SS	S	VSS	VSS	X	VSS	—
Zn⁺²	S	S	S	S	VSS	—	S	VSS	S	X	VSS	—
Fe⁺²	SS	S	S	S	SS	VSS	S	VSS	S	X	X	—
Fe⁺³	SS	S	S	S	—	—	S	—	S	X	X	—
Al⁺³	S	—	—	—	X	—	S	—	S	X	X	—
Ag⁺¹	—	S	SS	VSS	—	—	S	X	—	—	—	—
Pb⁺²	VSS	—	—	—	—	—	S	VSS	VSS	—	—	—
Hg⁺¹	—	S	S	—	—	VSS	S	—	VSS	—	VSS	—
Hg⁺²	SS	—	SS	—	—	—	S	S	VSS	—	SS	—
Cu⁺²	SS	S	SS	VSS	—	—	S	—	S	—	—	—

S -Soluble, over 5000 mg/l
SS -Slightly soluble, 2000-5000 mg/l
VSS -Very slightly soluble, 20-2000 mg/l
/X -Not a compound
X -Insoluble, less than 20 mg/l

Figure 48

Cause:
3. Biological

80%
Biologically Induced

Figure 49

Only the genus *Gallionella* has been shown to get energy from iron oxidation. Most of the typical iron deposition is not due to the typical filamentous forms of IRBs. Commonly, this diagnosis is based upon examination of a water sample under a light microscope. I do not trust the data from observations under a light microscope in determining the presence of IRBs. It can be useful as part of an analysis to consider: yes, we do have stalked IRB (*Gallionella*); yes, we do have filamentous IRB (*Crenothrix, Sphearotilus,* and *Leptothrix*); but do not use this technique exclusively (Figure 50).

It is important to use other techniques so that the full potential range of IRBs can be determined. Other tests such as the heterotrophic plate count, and the BART™ tests, are more useful indicators of biological iron deposition because they are going to give a lot greater range of the most common microorganisms that have the iron deposition capacity. If you are only using the light microscope, you are going to miss the true cause of most iron deposition. What we are looking for under a light microscope are these stalked (i.e., *Gallionella*), sheathed (e.g., *Crenothrix*) and filamentous bacterial forms which are relatively easy to identify with the microscope (Figure 51). For example, the species of *Gallionella* forms a typical twisted ribbon-like tail that is very easy to see under a light microscope. When the technician sees that then it can very easily become the focus. We have to shift away from the reliance on the light microscope to determine the IRBs. I have seen too many reports which say: "negative for IRBs," and when you look in the well from which the samples were taken, there is obvious evidence of biological filtration with iron deposition. This is due to quite a variety of other IRB species which cannot be identified directly by using light microscopy.

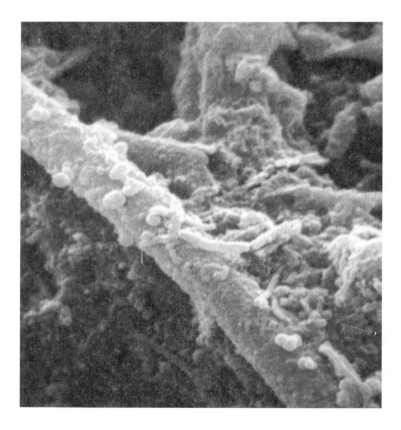

Figure 50

Figure 51

IRON-RELATED BACTERIA

The IRBs are normally considered nonpathogenic. They have been implicated sometimes in causing illness in weakened hosts; these are known as opportunistic or nosocomial pathogens. There are, however, a very wide variety of bacteria that are capable of iron deposition. It should be remembered that some of the total coliforms are also capable of iron deposition and these species can be considered as belonging to the iron-related bacterial group (Figure 52). The two genera of the total coliforms most commonly found are *Enterobacter* and *Citrobacter* that are part of the biofilm structures. When you pull a pump for repair or look down a well and there are iron deposits, it is quite common to find

Enterobacter and *Citrobacter* as parts of the bacterial flora being detected as causing the deposition.

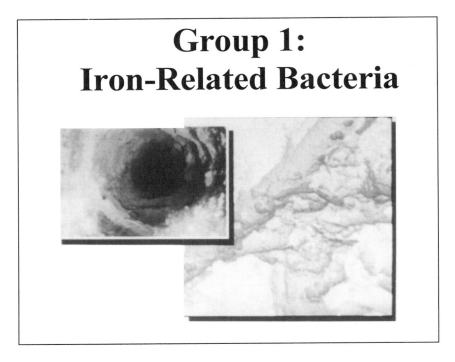

Group 1:
Iron-Related Bacteria

Figure 52

Sheaths (microbial tubes) are commonly seen when undertaking light microscopic investigations (Figure 51). Most of these sheaths are probably empty. There is some discussion among groundwater microbiologists as to precisely what is the function of these sheaths. A growing premise is that they are actually transport mechanisms to get oxygen and nutrients into the depths of the biofilm. If there is a surface and there is now a material buildup on that surface with minerals, biofilms, bacteria, and slime limiting the penetration of life supporting elements to the deeper zones, it is possible to see that these sheaths will actually penetrate into that biofilm. Where the sheath is empty, it essentially becomes a hollow tube which can function as a transport mechanism for oxygen and nutrients to the depths of the attached biofilms (Figure 53). The presence of sheaths can therefore mean that there is much more than sheath-forming bacteria present but also can indicate that there are many different biofilm formers that are growing deeper down, using the sheaths as transport conduits from the surface to the depths of the growths and deposit.

Figure 53

SLIME-FORMING BACTERIA

Most of the IRBs are slime-forming bacteria and most of the slime-forming bacteria are iron-related bacteria. (Figure 54) The common feature of slime formers is that they produce polymeric threads as protective slime capsules or coatings around the cells. That slime is necessary to get the bacteria to stick to the surfaces, and then it will produce extensive slimes as a protective coating. It is within that slime that most of the minerals are also trapped and/or become deposited.

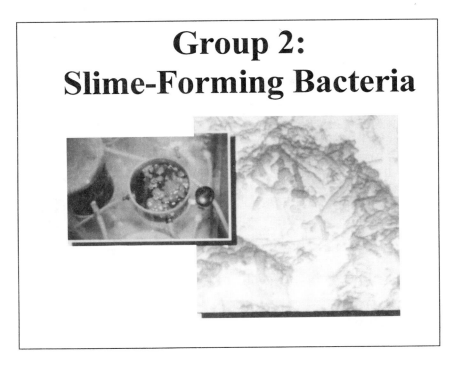

Figure 54

These bacteria are not independent groups but rather they are inter-related. Again I tend to focus less on actual bacterial names; I am more interested in what types of materials these bacteria are depositing or building up so that we can remove that material. Are they extensive slime formers? Are they capable of precipitating iron or manganese? Those are the types of questions that have to be addressed. The removal of all these materials is important in the rehabilitating of wells.

SULFATE-REDUCING BACTERIA

The sulfate-reducing bacteria (SRBs) will cause microbial corrosion by growing within a tubercle and under the other slime-forming bacteria in a biofilm or nodule. These SRBs will also generate the "rotten egg" or the hydrogen sulfide odors (Figure 55). This shows what happens when there are a number of zones created within a biofilm extending down from a surface through the biofilm to the attachment below. There is an aerobic zone closest to the surface. This aerobic zone perched in the top of the biofilm will lead to the reduction in oxygen due to metabolic activity. As a result, very little oxygen will pass right through this zone. This means that the zones underneath are anoxic (do not contain oxygen and so are anaerobic). This is where the SRBs will be able to thrive. Most of the time, these slime growths contain large and diverse bacterial communities. These bacteria are growing together or symbiotically, synergistically into the communities around one another. It is rare to find one microbial species all by itself!

Group 3:
Sulfate-Reducing Bacteria

- ❑ **Microbial Corrosion**

- ❑ **Hydrogen Sulfide Gas**

- ❑ **"Rotten Egg" Odor**

Figure 55

When we test for the various types of bacteria in ground water, we generally test for the total aerobic bacteria, for the IRBs, slime-forming bacteria and the SRBs. It is common to find most of those present most of the time! It is rare to find the absence of SRBs even if you are not having any hydrogen sulfide related problems (e.g., "rotten eggs," black waters and slimes, or corrosion). This is because these SRBs are growing, very much, with a whole community of bacteria which is becoming established within these water environments attached to surfaces in general.

10

DIAGNOSING WELL PROBLEMS

HISTORICAL RECORDS

When attempting to determine the cause of well problems (Figure 56), I tend to ask an almost endless number of questions! Historical records on the well itself are very important. If there is not any record of the original specific capacity to compare to, you really do not know what the target should be for the rehabilitation of a well — because you have no reference point. If you do not know your loss in specific capacity, you do not know how aggressive you should be in the rehabilitation of that particular well. For example, if you have a well that has only a 10% loss in specific capacity, you are not going to have to get nearly as aggressive as if you had a well which only has 10% of its original specific capacity (i.e., lost 90%). Here, you would have to get a lot more aggressive with this well because it is more severely plugged.

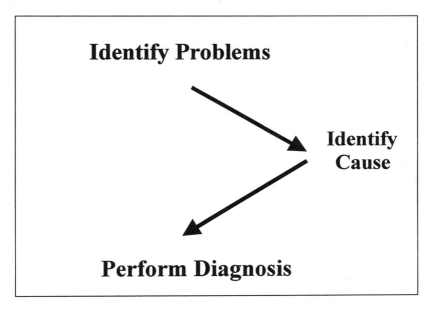

Figure 56

You have to have some understanding of just what is the status for a well and so it is essential to keep good records from day one. Not only is it important to have the historic specific capacities but it is also important to have the routine information on static water levels, pumping water levels and the rate of pumping. Again it must be stressed that it is most important to compare the specific capacity at comparable flow rate (Q) and static water levels.

This does not limit the information that should be gathered. Any information about the well needs to be considered. All of this helps in the designing of an appropriate strategy for treating that well. It is important to look at the water chemistry, bacteriology, down-hole video, and well logs (Figure 57).

Perform Diagnosis:

- □ **Water Chemistry**
- □ **Bacterial Testing**
- □ **Well Log**
- □ **Down-hole Video**

Figure 57

DIAGNOSIS UTILIZING WATER CHEMISTRY ANALYSIS

As far as the water chemistry is concerned, the most important part is to look at the mineral scaling potential (Figure 58). In other words, what are the minerals likely to precipitate from calcium to magnesium, from iron to manganese, total alkalinity, total hardness, pH, temperature, sulfates, phosphates and silicates? This now gives an idea of what minerals we are likely to have to deal with. In recent years, we have performed many deposit analyses. In other words, we will collect a sample of scale from the pump or the well screen.   This scale would then be analyzed to determine the constituents and also to determine the solubility characteristics. This will allow you to determine what chemistries are the most effective at dissolving that scale in a laboratory setting. By doing that, there is no longer as much guessing or educated assessment. We then have some practical knowledge of the most likely treatment scenario based upon these laboratory studies.

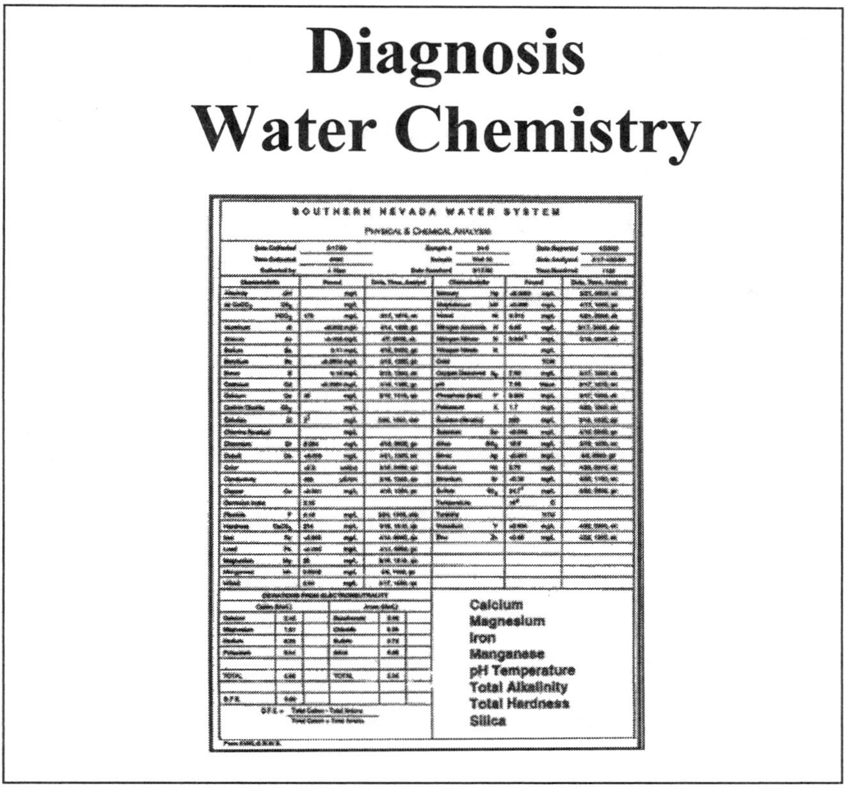

Figure 58

I will caution, however, against over-interpretation. Remember that the study on one sample of scale may be in error since that scale may not be typical of the scale found elsewhere in that well. While this sample might not be totally representative of all of the scale deposits that might exist in the well, this information is taking you one step closer to an appropriate treatment selection rather than just using indirect water chemistry parameters to make that determination on the scaling potential.

DIAGNOSIS WITH DEPOSIT ANALYSIS

In using the laboratory scale analysis, it is possible to be much more successful, and I will give an example. In Texas, we did a well rehabilitation with good chemistry and what appeared to be an effective treatment and got no improvement in specific capacity, and the well looked the same during a video inspection. When we went back to the deposit analysis, we found that this particular well had phosphate-based deposits. Phosphate often forms very hard types of mineral scale. These are very difficult to dissolve. Even with powerful chemistries, we were only able to get 18% into solution. Now the limitations for this well become very clear — it is finding the means to get these (very hard) deposits to dissolve. We ended up with this well going back with the same chemistry but very much more aggressive mechanical agitation and, even though we had only dissolved 18%, we managed to destabilize the structure enough to get it to collapse once enough energy was applied to it. Once you come to understand the challenges and limitations of trying to get the scale to dissolve, then the importance of doing the deposit analysis becomes clear. You replace an uneducated guess with information about the scale in that particular well.

DIAGNOSIS WITH BIOLOGICAL ACTIVITY REACTION TESTS (BART ™)

As well as the chemistry, we also examine the bacteria that are fouling the well. Because of the significance of each of these groups (IRB, SRBs, slime formers and heterotrophs), these are the ones commonly tested for in groundwater or water well situations (Figure 59). We've used BARTs now for many years in the U.S. These are developed and manufactured by Droycon Bioconcepts Inc. here in Regina. They are excellent tests (Figure 60 and 61). You can get a better assessment of groundwater microorganisms with a BART™ than you can with a heterotrophic plate count or microscopic analysis. The advantage of the BART™ is that you are able to create enough differing environmental conditions that allow you to be able to grow a wider range of bacteria than are found in that particular groundwater or water well sample. If you start to put bacteria that are accustomed to growing in a well onto the semidry surface of an agar plate to incubate, you probably are not going to be able to grow many of them.   This means

Diagnosis:

Bacterial Testing

- Iron-related bacteria (IRB)

- Slime-forming bacteria (SLYM)

- Sulfate-reducing bacteria (SRB)

Figure 59

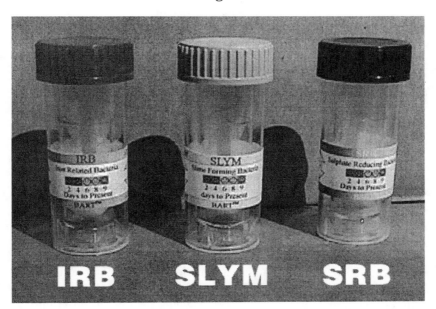

Figure 60

Figure 61

that the colony count from that agar plate is going to be a gross under-estimation of the numbers of bacteria actually in the water sample because you have not been able to grow them to form visible colonies that can be counted. The bacteria from the water are being placed into a foreign environment where they are being stressed and they are not able to overcome that in time on the agar surfaces. With these particular (BART^{TM}) tests, you can overcome some of that problem by creating a whole series of different environmental conditions where these bacteria can grow and be assessed. These differing environmental conditions are created in part by a ball, which now floats up to the surface of the water sample. When 15 ml of water sample is added, the ball floats up to the surface and restricts oxygen entry. This creates the different aerobic to anaerobic, oxic to anoxic or oxidative to reductive conditions from the top to the bottom of the tube, respectively (Figure 62). This allows a gradient of environmental conditions to be created in the tube. A second gradient is created by the selective nutrients. These are dried as a pellet into the bottom of the tube. When the water is added, these nutrients begin to dissolve and form a nutrient gradient front moving up the tube. The bacteria in the sample are thus presented with a whole range of different environmental conditions into which they could grow only if the nutrients are selective for their growth.

Remember that a lot of the bacteria in groundwater are not used to high nutrient levels, so much so that they are organo-sensitive (will be

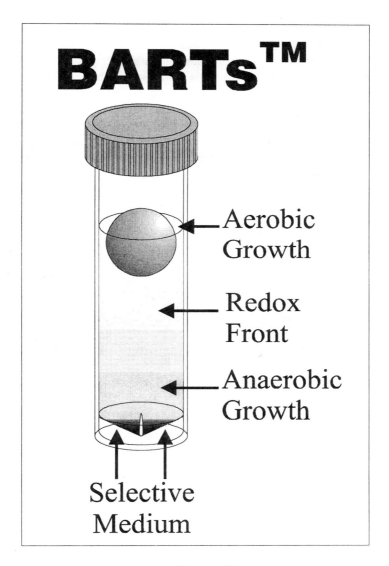

Figure 62

traumatized by too much nutrients). Putting these bacteria into a BART™ allows them time to adapt to the increasing nutrient levels coming up on the gradient front. Thus the nutrients go from a high concentration at the bottom to a low concentration at the top. This cannot happen on an agar plate where the bacteria are subjected to stresses from too high a nutrient concentration and too little available water to allow them to survive and flourish. Sometimes in the BART™, the zone where growth begins to occur can be very narrow, like a slimy cloudy plate floating in the water under the right environmental conditions for bacterial growth. This demonstrates the principle that you are going to encourage the bacteria to grow in the tube under conditions favorable for their growth rather than favorable for the technician who wants to be able to count colonies!

These BART™ tests can be both semiquantitative and semi-qualitative. Information can be gathered by observing the number of days of delay (or time lag) for the start of growth. This is indicative of the number of bacteria present, the population size or the aggressivity of the population in determining whether there are problems or not (Figures 63, 64 and 65).

Biological Activity Reaction Tests
(BART™)
A simple test for bacteria in water

□ **Add water sample up to the fill line (15 ml)**
□ **Place the test at room temperature**
□ **Observe daily for any reactions**
□ **Note the time lag (days of delay) to each reaction**
□ **Determine aggressivity based on time lag**
□ **Determine bacterial type by reaction(s) observed**

Figure 63

Again, I tend not to focus so much on numbers as on ranges. Is it a significant problem, is it a minor problem, or is it not a problem at all? Instead of focusing on "we have exactly 333.5 bacteria in each mL of water" — not that important — it is much more important to focus on the ranges if it is a particular problem. You can, by comparing against

Iron-Related Bacteria (IRB)

Reaction Pattern Codes:

- ❑ BC - Brown Cloudy (reaction 4)
- ❑ BG - Brown Gel (reaction 3)
- ❑ BL - Black Liquid (reaction 10)
- ❑ BR - Brown Ring (reaction 4)
- ❑ CL - Cloudy Growth (reaction 2)
- ❑ FO - Foam (reaction 5)
- ❑ GC - Green Cloudy (reactions 8 and 9)
- ❑ RC - Red Cloudy (reaction 7)

Figure 64

Sulfate-Reducing Bacteria (SRB)

Reaction Pattern Codes:

- ❑ BB - Blackening Base (reaction 1)
- ❑ BT - Blackening Top
  around the ball (reaction 2)
- ❑ BA - Blackening All Over (reaction 3)
- ❑ CG - Cloudy Gels (reaction 4)
  (CG is a negative detection for SRB)

Figure 65

common reaction patterns that are seen in the BART™ tests, determine what types of microorganisms are growing down in the well and what types of environmental conditions are occurring. It may also be possible to identify particular groups of bacteria. For example, one of the color reactions may be caused by *Enterobacter*, while another might be caused by *Citrobacter*. They are excellent screening techniques for wells. The SRBs will produce the black deposits of iron sulfides. They are easy to read since the test blackens at the base of the tube (reaction one) or around the ball (reaction two).

DIAGNOSIS USING HETERTROPHIC PLATE COUNT

We have had a lot of heterotrophic plate counts (HPC) done over the years, commonly in conjunction with the BART testing. Here, we will usually do timed interval sampling from the well being pumped. Here, it is advantageous to let the well "rest" (i.e., not be active) for about twelve hours (Figures 66, 67 and 68). By allowing the well to rest in a nonpumping condition, you are causing a greater detachment of bacteria from the biofilms and that means that you are going to get these additional microorganisms to be pumped with the initial discharge of water from that well when it is activated again. It is sometimes possible to determine, by collecting samples at 5, 15, 30 and 60 minutes after a well has come on, a difference in the microbial populations associated with the detachment of microorganisms from those biofilms around the well. I believe that these heavier burdens of bacteria found in water samples collected in sequences reflect mainly detachment (shearing) from the biofilms in the well, although there are some researchers who suggest that the distance of those growths from the well can also be estimated. I do not believe that you can get so accurate most of the time.

Heterotrophic Plate Counts				
Well location	5 mins	15 mins	30 mins	60 mins
Enterprise, AL	25	0	4	0
Hartford, AL	TNTC	30	3	4
St. Mary's, LA	37	40	22	21
Baton Rouge, LA	280,000	150,000	600	3

Figure 66

Heterotrophic Plate Counts

Well location	5 mins	15 mins	30 mins	60 mins
Gordon, NE	0	510	75	50
Pleasanton, NE	313	112	140	55
Jamesville, WI	0	0	1	116
Olin, IA	140	71	36	3
	33	24	9	0
	410	235	500	0
	205	75	0	0

Figure 67

Heterotrophic Plate Counts

Well location	5 min	15 min	30 min	60 min
Louisville, KY	10	1,880	2,440	10
Athens, GA	4,400	2,140	1,800	560
Elkhart, IN	0	2,600	0	0
Peru, IN	19	32	110	2

Figure 68

You cannot say that particular bacterial fouling can be occurring at a specific distance from a well (e.g., 8' 2"), but rather that you can gather some general information about the types of bacteria fouling in the zones immediately around the well. The emphasis has to be on finding the bacteria detaching from surfaces. The fact that the SRBs may not appear until the 30-minute sample could mean that the SRBs are deeper in the biofilms and so cannot detach into the water until the upper layers (of aerobic bacteria) have been stripped away by the pumping action of the well. There needs to be caution in the interpretation of the data as to whether these bacteria are being pulled in from further and further into the formations or whether the bacteria are continuously detaching from the same biofilms. Decreases in the numbers or aggressivity of bacteria in timed interval samples may be a simple reflection of the slowing down in the rate of detachment rather than reflecting the arrival of bacteria from greater distances. It is not possible to categorically link the delay in the appearance of particular bacteria in the timed pump sample with a location prediction for its origin.

DRILLER'S LOG FOR DIAGNOSIS OF PROBLEMS

In addition to the history, chemistry and bacteriology of the well, it is also important to look at the driller's classification (and/or e-log) for the well (Figure 69). It is important to know what formation the well is in to examine the potential for mechanical type blockage. This is the potential for particulate material being part of the blockage.

Diagnosis:

Well Log

❑   **Electrical Classification**

❑   **Electrical Log (E-log)**

Figure 69

DOWN-HOLE VIDEO INSPECTION

Down-hole videos are extremely useful tools. Until the entry of the down-hole video inspection, we really did not have the ability to see inside of a well (Figure 70). It was always completely out of sight. Now there have been thousands of wells that have been video inspected. It is very useful to determine not only the wells' structural failures (such as holes in screens and casings), but also to get some idea as to the coloration and also some idea as to what needs to be done for rehabilitation. Is it going to be important to pre-clean a well before chemical application, along with wire brushing, swabbing or Sonarjet™? I have seen videos of wells that have been significantly off in their specific capacity and yet you take a look at that well and it is relatively clean on the inside! That means that you would be wasting time to try and clean the inside of the well because it is not badly fouled. It therefore gives you an idea of where the fouling is (in this case, behind the well screen) and you can concentrate on plugging deposit removal that exists beyond the well screen.

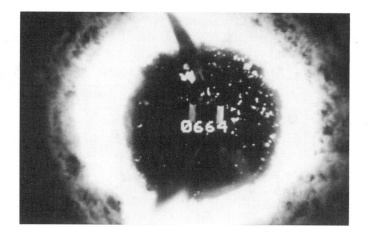

Figure 70

11

WATER WELL REHABILITATION

INTRODUCTION TO SOLUTIONS

There are many different strategies that can be used in water well rehabilitation. Most of my time is spent on assessing well problems and making recommendations (Figure 71). These are many different techniques that we are currently using because there is not "one size fits all" (Figure 72). Some of the techniques used for rehabilitation (such as Aqua Freed™) are relatively broad range in their ability to disrupt and remove a wide variety of deposits. Some of the chemistries, such as QC-21 Well Cleaner™ are also relatively broad range in the type of deposits they will dissolve and disrupt.

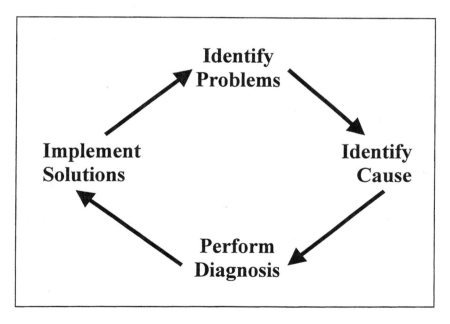

Figure 71

Rehabilitation Treatments

- ❑ **Can involve many different strategies**
- ❑ **Must achieve effective deposit removal**
- ❑ **Must be custom tailored, based upon cause of problem, well construction details, and type of formation**
- ❑ **Must have penetration into the surrounding formation**
- ❑ **Must have good agitation**

Figure 72

The focus here will be on the techniques that I have the most confidence with, based on experience with most well rehabilitation procedures tried or utilized. We will discuss some of the important aspects of choice of chemistry, and the application of chemistry with redevelopment techniques. Just one more time, it has to be remembered that the BOTTOM LINE is EFFECTIVE DEPOSIT REMOVAL.

As discussed earlier, this gets down to the longevity issue, and initially better results of restoring capacity and solving water quality problems. This means the increasing of the time frame between treatments by getting back to the original condition of the well. Even though a well rehabilitation treatment indicates that the specific capacity is one hundred percent of original, all the material may not be removed. It is important to get back to the original surfaces so that you have that excess production capacity. To do this, you have to clean those surfaces so that you do not leave any organic debris material behind from which new growth can immediately get a "boost".

Rehabilitation treatments need to be custom tailored to fit the well that is being treated, bearing in mind the cause of the problem, the well construction details (types of screen openings, size, gravel pack, the type of aquifer materials and formation as well). You must custom tailor the the treatment appropriate to these details. Again, historically with regards to water well rehabilitation, that was not often done. It is surprising how many contractors will take the approach "One Size Fits All!" They will take the approach of acidization, for example. In that case, every well problem is going to be solved by acidization. In other cases, the contractor may believe that all of the problems will be solved by chlorination. This is just not going to work! The reason for this is that natural variations that are occurring in the types of problems that cause wells to begin to fail. Understanding the specific cause of these variations for each well is essential in determining the appropriate treatment for that given well.

One of the most basic and fundamental aspects of the treatment is that you have to have good agitation (Figure 72). It does matter how good the chemistry is, but the more important aspect is to get good agitation in order to break the material up. The chemistry is important to soften this material up and dislodge some of that material from surfaces but it will not bring everything into solution. That is why agitation is so essential; it is to achieve a "washing" action in and out of the formation.

Chemical treatment under a static (stagnant) condition is just not going to work most of the time without agitation to improve the chemical action. Every time there is agitation by surging, by brushing or by swabbing, there is more contact between the chemicals and the deposits and there would be better cleaning of the surfaces and more complete removal of those deposits. In recent years, it has become clear to me that water well rehabilitation can be separated into three distinct steps (Figure 73).

Steps for Effective Well Rehabilitation

□ **Pretreatment**

□ **Various Treatment Applications**

□ **Development or Redevelopment**

Figure 73

PRETREATMENT

This means that you clean the regions of the well that are easy to get at (Figure 74). These areas of the well that are easy to get at are the inside of the well screen or the face of an open hole. Historically, there have been treatments that have been limited to just wire brushing a well. California was such a state where that was done. When I first joined Layne nine years ago and visited sites in California, I was told that the normal practice for water well rehabilitation was to brush and bail the well! That was all they did. Basically just scratching the inside of the well is literally just scratching the surface of the problem. In recent years, much more effective treatments are now being employed. Having said that, it has to be remembered that this stage is important to remove the material that is easy to get at. This means that there is now a higher confidence that the application of chemicals or carbon dioxide later is more even and more likely to penetrate beyond the well screen. This step also ensures that the chemicals have a greater chance of reaching the deposits that do lie on the other side of the well screen. Also, removing material that is easy to get at mechanically makes it easier for the chemicals to attack and dissolve the underlying materials.

Pretreatment

- ❑ **Can include – wire brushing, swabbing, airlifting, Sonar-Jet^{TM}, etc.**
- ❑ **Removes material primarily from inside the well**
- ❑ **Ensures more even application of chemicals or carbon dioxide**
- ❑ **Follows the "path of least resistance"**

Figure 74

VARIOUS TREATMENT APPLICATIONS

After pretreatment is done, the next stage is going to involve treatment applications, usually with chemicals and/or liquid carbon dioxide. This now forms the bulk of the treatment which takes place after the cleaning of the inside of the well. This step is designed to penetrate the gravel pack and the formation and disrupt the material that lies beyond the well screen. The most important component of this step is to achieve the detachment of material from those surfaces of the well screen, gravel pack (if gravel packed) and the surrounding formation.

DEVELOPMENT OR REDEVELOPMENT STAGE

This involves removal of the sludge and removal of the deposits after they have become detached from the surfaces. Sometimes all of this is attempted with the pump in place and people still expect that by over-pumping the well, all of the sludge and deposits will be removed from the well. So often this stage is neglected, and yet it remains the crucial final stage in water well rehabilitation. One of the techniques that are commonly used involves the technique called over-pumping.

If you have a well that has a design capacity of 1,000 gpm and you will temporarily pump that well at 1,500 gpm for development purposes, the idea is to over-pump the well to get high velocities to lift and get removal of the sediment and sludge. However, it is probable that you are not going to achieve that, since you may not have sufficient velocities necessary both in the surrounding formations to get the migration of the particles into the well and the up-hole velocities to remove the particulate material from inside the well. There is not enough concentrated energy to do this. One of the techniques that I will be describing involves just this technique for which a large database is now building. Here, the objective is to focus on small sections (4' to 5' at a time) of the well screen at one time and then apply the energy to those sites only. If you can imagine doing this for say a 700' well screen, it takes a longer period of time for development but you should be able to expect more complete removal of deposits. By concentrating on small sections of a well screen, we are able to measure markedly improved specific capacities. Again this is because there is a much greater energy level with higher velocities being applied to shorter sections of the well screen at one time.

Pretreatment is easy to do. In Layne, we often make our own wire brushes; it's easy to do. The manner in which some of the brushes are normally made is to take a pipe, drill some holes through the pipe and pack these holes with wire rope and fray the ends so that they will scrape against the sides of the well when being applied. Other brushes look more like street sweepers but all scrape the inside of the well. Many different types of well screens are wire brushed but may require some special consideration. There has been some concern with wire brushing wire-wrap well screens and causing damage to the screen. Wire-wrap well screens are often wire brushed in many parts of the U.S. If all of the materials that were plugging the well were on the inside of the well, then this would be a very effective method of rehabilitation with a 100% success guarantee simply by scraping that material off. It is a lot more difficult to disrupt and remove plugging materials that are beyond the well screen because they are in inaccessible areas for the most part and are more difficult to get at.

Swabbing resembles using a plunger (Figure 75). A swab is placed down on the steel pipe and consists of a couple of bolted steel plates. Between the plates is a rubber-belted material and it is this rubber belting which makes contact with the inside of the well screen. We will often have two of these swabs separated by approximately five to ten feet of pipe with holes going down into the producing zone of a well. Chemicals can be injected between these swabs (Figure 75).

In some states, such as Texas and California, most of the time we will not treat a well without injecting the chemical this way. We are injecting the chemical through approximately every five feet of well screen even when the screens are hundreds of feet long. This same approach can also be used on shorter lengths of well screen that are only 20' or 50' long with the same benefit. This method of chemical injection means that every section of a producing zone is more effectively treated to follow the pathway of least resistance through the screen, gravel pack and into the aquifer material. The injection of the chemical through a relatively small part of the well screen tends to assure that the chemical will be more evenly applied. After the chemicals are injected, the whole swabbing assembly is gradually moved up and down the well in each section of the screen to achieve a surging action and agitation for the maximum impact. The swabs here are creating differential pressures, and so you have this swab cleaning off the inside of the well while at the same time creating a positive pressure on top pushing chemicals into the formation and a negative pressure underneath pulling disrupted sludge and deposits in from the formation. All of this cleaning activity is generated by the application of differential pressures. Even though agitation is one of the important steps of chemical application, it is possible to agitate too much when the chemicals are first put in the well.

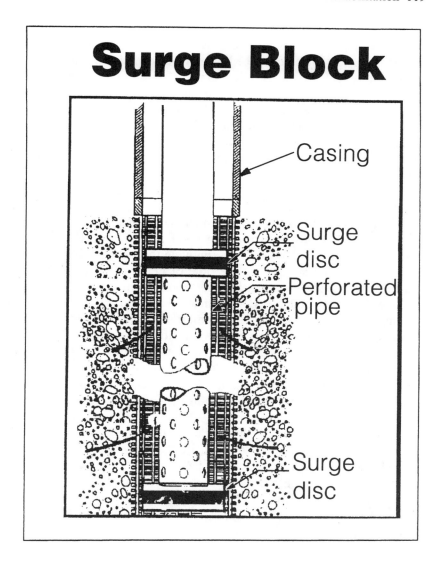

Surge Block

Figure 75

A calculated volume of chemical with a strong enough concentration is placed into a well. It is very important to have strong enough chemicals in the zone close to the well screen or for a short distance into the formation. The danger of swabbing with chemicals too much when the chemical is first placed in the well is excessive dilution of the chemical to a point where it would no longer be effective.

Air lifting can be used in conjunction with swabbing also for pre-treatment or as will be described later for development. Air lifting, as such, simply refers to the use of air to pump water. In the simplest form, an air lift pump is a pipe with the air line terminating within that (inductor) pipe. When air is applied, the air can lift and pump thousands of gallons of water per minute relatively easily. Again, it is possible to isolate sections of the well in order to maximize the localized treatment. This is really designed to take out the material that is easy to remove and has already been disrupted or detached from surfaces.

SURGE AND PURGE

In recent years Layne has performed many wells with a "surge and purge" procedure, where alternating application of gases is used for disruption of material and also for removing the material from the well by lifting it to the surface. At the time of publication, this process is currently being filed for patent. The application of gaseous carbon dioxide (or other suitable gases) under pressure is used to force a column of water with or without chemicals rapidly into and out of the formation. This can be very effective in the detachment of material. Once the material has been detached from the surfaces in the well environment, it can be brought out of the well with gaseous removal or submersible pumps placed in the development tools. This procedure has been used as a pretreatment and also as the only process used for well rehabilitation with very good results.

The results evaluated from the database show significantly improved results compared to some of the other well rehabilitation procedures. This process focuses on the development issues, which are often not given enough attention. It should be remembered that when the well has been pretreated to remove some of the plugging material, it means that chemicals are not going to be spent dissolving deposits or disrupting biofilms that can be easily removed from the well.

SONAR-JET™ WELL REHABILITATION

Sonar-Jet™ is another technique that I have had quite a familiarity with in treating thousands of wells (Figure 76). We have used it in many parts of the country. In some cases, that was the only treatment that was applied. Sonar-Jet™ is a vibratory explosive using an expansive gas.

We set off a series of charges in a sequence so that a series of vibrations are created down-hole. It is a very effective method for cleaning the material off on the inside of a well screen and for a short distance into the surrounding formation.

Solution:
Sonar-Jet™

Enhances Chemical Treatment

Figure 76

It has been found that the Sonar-Jet™ has not been damaging to wells and the major cleaning effect was on the inside of the well screen, outside the well screen and for a short distance into the gravel pack (Figures 76 and 77). The size of the charges is based upon the size of the well, (e.g., diameter) and submergence of water and is such that, while shock waves are created by the explosive detonations, it is not enough to damage the well. This is often used as a pretreatment and it has been found to be good for that purpose. Often the Sonar-Jet™ equipment is in the video inspection van and it is common practice for the well to be video-logged, and if shown to be plugged, immediately treated with the Sonar-Jet™ (Figure 78).

CHEMICAL APPLICATIONS IN WELL REHABILITATION
The application of chemicals has long been a part of water well rehabilitation (Figure 79). Acidization and chlorination of wells has been used for decades. There are many occasions when the chemicals used have not been properly selected or properly applied (Figure 80).

Sonar-Jet™ Benefits

❏ **Removes Scale from Screen or Open Hole**

❏ **Restores Wells Connection to Formation**

❏ **Reduces Chemical Demand**

Figure 77

The important aspect, when it comes to the chemical treatment of a well in general, involves the twin parameters of volume and concentration (Figure 80). These are the two most important considerations although it may sound simplistic. It remains surprising that many people will treat a well without even doing a calculation to determine the volume of the well! Again, there is that attitude of "one size fits all" and, here, it is thought that you do not need to worry about the volume and the concentration.

For example, let's do a 1,000 gallon acid treatment. Well, what does that mean if you do not know the volume of the well and the final concentration of the acid when it becomes diluted? There have been occasions when individuals familiar with well rehabilitation have moved from one area where the wells are 50' deep to an area where the wells are 1,000' deep and think that the same treatment is going to work even though there is an obvious difference in scale. Clearly, the volume of chemical to be used has to be related to the volume of the well if you want to displace and get some penetration into the surrounding formation.

It must be remembered that the volume you would like to displace with the chemical is not simply the volume of the well itself but also those volumes within the surrounding formations for that distance which you would like to treat with the chemical. Commonly, it is better to aim to treat twice or three times the well volume in order to ensure adequate penetration of the surrounding formations with the chemical.

Figure 78

Figure 79

Solution: Chemical Treatment

Site-Specific Customization and

Chemical Treatment Considerations

- ❑ **Volume**
- ❑ **Concentration**
- ❑ **Sequence**
- ❑ **Combination**
- ❑ **Above-ground premixing**

Figure 80

The next consideration is that the chemical is strong enough, that there is enough concentration, to dissolve the minerals, scales or deposits. One thing that automatically is being faced is the dilution of the chemicals in the well with the well water. There are two events happening. First, the chemical is displacing water in the well back into the formation. Second, the water is mixing with and diluting the chemical. As a result, the application of the chemical is being affected by the interactions between the displacement and the dilution effects. Essentially, displacement is going to be a major factor in the well itself. For these reasons, applying concentrated chemicals directly to the well is not suggested and it is very important to premix chemicals above ground. Outside of the well beyond the well screen, it will be the dilution factor that now dominates. This is because there is such a large volume of water out there in the gravel pack and formation that it will quickly dilute out the chemical. Clearly, as the chemical moves further away from the well screen, the concentration will fall rapidly to below the effective range. This is likely to happen within a matter of feet. The limitation of the treatment zones will, in all probability, be this dilution effect. As a rule of thumb, every time that the distance or diameter is doubled there is a corresponding four-fold increase in the dilution.

Above-ground premixing of the chemicals is also very important. You cannot take a concentrated chemical such as an acid or chlorine and simply dump it into a well. If this is done, then you end up with a mixture of zones. Some zones would be super-hot with a lot of chemistry, while other zones would have hardly any action. That is not efficient. If you were, for example, to dump an acid down a well, there may be a "flood" in one zone with very low pH values reached while other zones would be buffered away from the "flood" and would have a very small pH shift downwards. It is for these reasons that it is better to mix the chemical in tanks above ground so that there is a better and more even application down-hole. The mix can then be introduced into the well to cause displacement and ensure a more even mixing. As previously described it is even better to inject into every zone (Figure 75). There is no advantage to high pressure injection of chemicals and the high pressure injection may actually be negative. It may be necessary to overcompensate the concentration of chemicals. This over-compensation is necessary because of the dilution effect of the chemistry with the formation water. The final concentration is the important aspect in chemical application. As mentioned earlier, agitation when the chemical is first placed in the well can lead to further dilution because of the increased exchange between formation water and the chemicals placed in the well.

Hydrochloric acid is the number one choice for acidization. The concerns about corrosion due to the acid can be controlled by the use of inhibitors. If you use the uninhibited acid, then there is a corrosion risk. There are a number of different inhibitor products such as Rodine, for example. The Rodine products used in potable water are Rodine 100-AquaHib, which is claimed to be safe in potable water, and Rodine-103. There are also a variety of wetting agents (surfactants), amines that can be used for corrosion inhibition. The one product that we commonly use passivates the surface to inhibit the risks of corrosion. That passivation would only occur on steel surfaces and the ferric deposits would not become passivated which would control corrosion without losing the advantages of hydrochloric acid dissolving the scale.

There is no doubt that the uninhibited hydrochloric acid is very corrosive, and many people have shifted away from this acid for those reasons toward other acids. Alternative acids include hydroxyacetic, citric, acetic, sulfamic, and oxalic. These are all less corrosive but they loose the ability to be able to dissolve the minerals. What determines the ability of an acid is the availability of the H^+ ion. This is referred to as the ionization constant and will be described later. Even though the pH of the solution may be low for hydroxyacetic acid it is often used simply due to the fact that it is less corrosive. It is a mild organic acid and is less corrosive. However, if we compare the ionization constant, it is approximately 826,000,000 times less effective at dissolving mineral deposits than hydrochloric acid, because it will not ionize completely to provide the hydrogen ion. Hydrochloric acid remains the best chemical of choice because the corrosivity can be effectively controlled and its ionization is immediate.

PHOSPHORUS TREATMENT PRODUCTS

Another aspect of the Department of Energy study was that they determined the nutritional status of the microorganisms that they recovered from the well cores. The limiting nutrient was phosphorus and yet what are the chemicals that are often used in well rehabilitation? These are phosphorus based, such as phosphates. I would like to caution people against using phosphates. There are products on the market which are being used in wells to take care of bacteriological problems. The use of a limiting nutrient to take care of bacterial problems is like fighting fire with gasoline. These phosphate-based products have been utilized recently for "unsafe" bacterial problems. I have had the opportunity to recommend alternate solutions to bacterial problems experienced on wells after phosphates were utilized. It is my opinion that these results would be short lived and would not likely take care of problems long term.

We have to leave phosphorus out of wells in general because it is the limiting nutrient for microbial activity and adding it to a well makes it no longer a limiting factor, and there can be an explosion of microbial growth. It is good for the development of wells and the dispersion of clays but it is very important to use good disinfection procedures after the application of phosphates for mud and clay removal. Some alternatives may be used which are not going to stimulate the growth of bacteria in the wells but at this point in time have not had enough testing to see if they work as effectively as SAPP in the removal of muds and clay.

Whenever phosphorus-based treatment chemicals are added to wells, this literally "feeds" the bacteria in the system because the phosphorus is no longer the limiting nutrient – the nutrient that was restricting growth is now encouraging it. One downstream effect of the use of phosphates is that treatment times gradually get shorter and shorter as the added phosphorus with treatment stimulates the bacteria more and more. It becomes a vicious circle: treat — control — feed — grow and the phosphorus builds up in the microbes around the well! The end result is shorter time frames between plugging. As described earlier in the longevity section, the phosphorous is not the only reason for the shorter time frame between treatments.

SHOCK CHLORINATION AND OTHER OXIDIZING AGENTS

I believe we rely too much on oxidizing chemical treatments to remove material from wells. For example, shock chlorination is limited at the best of times as a treatment. Shock chlorination has to be one of the most inadequate procedures that we use for restoring capacity, but it is still a very useful part of an overall treatment process. Shock chlorination does not remove material, it does not dissolve material. Shock chlorination (chlorine gas, sodium hypochlorite, calcium hypochlorite), chlorine dioxide or potassium permanganate, etc. are oxidizing agents, and are actually going to do the opposite. They are going to be effective at oxidizing organic material to get some slime dispersion. But oxidation of the dissolved minerals in water would lead to additional deposition of minerals. The bulk of the deposited material (normally 75 to 80%) is mineral in content and would not be removed with chlorine or oxidizing agents. These oxidizing disinfectants are going to take care of the softer material but the bulk mass which is more mineral in content will not be affected. If you are trying to restore capacity using shock chlorination or chlorine dioxide, it is most likely not going to work most of the time, depending upon what the deposits consist of. Whatever can be done to remove the bulk of the mineral deposits, that is the way to recover specific capacities in a well. Shock chlorination alone will normally not be able to do that. Whatever can be done to get effective deposit removal is only going to be beneficial, and you can start reversing in a shorter and shorter time frame between treatments. We often expect that something as simple as shock chlorination will work, but it is just not capable of removing the bulk mineral deposits which form the majority of the problems with lost capacity.

DISINFECTANTS

Of the disinfectants that are used, sodium hypochlorite is about the best disinfectant that we can use (Figure 81). First of all, you cannot get high enough concentrations of chlorine if chlorine gas is used. Not only that, but the chlorine gas is also difficult and potentially dangerous to handle. Given the risks with chlorine gas, the choice comes down to using the calcium or sodium salts of hypochlorite. Sodium hypochlorite is the preferred choice over calcium hypochlorite. You can get 12% industrial strength or just over 5% for the domestic bleaches. The calcium hypochlorite has 65% available chlorine. However, the sodium hypochlorite is actually cheaper per unit of available chlorine. In addition to that, another disadvantage of the calcium hypochlorite in comparison to the sodium salt relates to the application of the chemical. Whenever you put these salts of hypochlorite into solutions, they will separate ionically into the water.

Step 1: Disinfectants

- ❑ **Sodium Hypochlorite (bleach)**
- ❑ **Calcium Hypochlorite (HTH)**
- ❑ **Chlorine Gas**
- ❑ **Chlorine Dioxide**
- ❑ **Sodium Hydroxide (lye)**

Figure 81

Laboratory studies have shown that the calcium hypochlorite is 50% less effective as a disinfectant than the sodium hypochlorite. These results were performed under carefully controlled conditions. It is thought that the calcium acts as a binding agent for bacteria; it also tends to get fixed on the terminal ends of the polymeric strands that are commonly produced by bacteria. The calcium therefore becomes a problem due to this reactivity. In addition, if you are dealing with water that contains bicarbonates, you have the potential for a lot of calcium carbonate deposits in the water. If you have ever dissolved calcium hypochlorite in water and you see a lot of white flaky "stuff", that is exactly what you are doing — you are generating calcium carbonate precipitates. This can be a very negative event. If you are doing this down-hole and are not able to flush these precipitates out effectively, these precipitates can collect and scale up the well.

The better choice as far as the disinfectant goes is sodium hypochlorite (otherwise known as liquid bleach). In addition to that, this is also very important when looking at effective well disinfection. If you are concerned about an unsafe bacterial sample, chlorination is often not just the only thing that needs to be done. We often have to take exactly the same approach with this unsafe bacterial sample as we would for the well if it had lost specific capacity. Both of these needs involve removing the same material! Although the goals for disinfecting the well would appear different from bringing back a lost capacity, rehabilitation of the well in removing the buildup of biofilms, scales and deposits achieves both of these goals. Disinfection will not dissolve the deposits harboring bacteria and will not bring back capacity, but a full rehabilitation has the potential to remove those bacteria along with the other deposits and essentially "disinfect" that well. In other words, an unsafe bacterial sample may not be tackled simply by the application of a disinfectant alone because many of the bacteria may be protected by the various biofilms and deposits in the well system. We therefore have to become more aggressive in dealing with water well problems to ensure that all of the deposits, scaling, nutrients, biofilms and tubercles are dissolved, disrupted and removed. Only then is it probable that the well would not be experiencing an unsafe bacterial problem. Better chemistries are clearly a major factor in the rehabilitation and disinfection of these plugged wells.

Disinfection, when using hypochlorite, involves the killing of the bacteria due to the production of hypochlorous acid ($HOCl$), which is the active form of the disinfectant. If we are putting sodium or calcium hypochlorite into water, hypochlorous acid is automatically formed depending upon the pH of the water. At the same time the hypochlorite ion is also being formed, depending upon the pH. When the pH falls below 7 in water, then there are more hydrogen ions being formed than hydroxide ions and these start to change from the hypochlorite ions to become hypochlorous acid. When the pH rises to the more alkaline end above 7.4, then the conversion shifts to the hypochlorite ion as it looses hydrogen ions. This hypochlorite ion is not nearly as effective as a disinfectant as the hypochlorous acid is. For example, if the pH of the water is at 8.4, 95% of the chlorine is automatically converted to the ionically charged hypochlorite ions. This charged form interferes with the transport of the chlorine into the bacterial cell. What is happening here is that the charges on the surface of the bacterial cells are such that, under these more alkaline conditions, the chlorine is no longer able to penetrate as readily as an uncharged molecule into the bacterial cell.

There is an attraction between different charges resulting in inhibition of transport to the bacterial cell. I generally set an upper limit of a pH at 8.0 to consider either lowering the pH or selection of alternate disinfectants. At higher pH values, the hypochlorite cannot be effective and some alternative disinfection strategy needs to be considered. One method is to actually lower the pH with acids to improve the effectiveness of the hypochlorite. The acid that would be used here would not be hydrochloric acid. This would be dangerous because the hydrochloric acid could start reacting with the hypochlorite to form mustard gas (chlorine gas). One acid that could be reasonably safely used would be sulfamic acid. Even lowering the pH just a couple of units below 8.0 can prove to be very effective prior to the actual disinfection. It has to be remembered that pH can have a very significant impact on disinfection efficiency and lowering the pH has to become a part of the strategies for the disinfection of the more alkaline waters with a pH of above 8.0.

Under some circumstances, such as pH limitations, it is possible to go to other chlorine-based disinfectants (Figure 81). One of these disinfectants sometimes used is chlorine dioxide. Here, the chlorine dioxide has three times the disinfection capacity and is not as pH sensitive as hypochlorites. Generally chlorine dioxide cannot be generated at-site although it can be generated in the well by utilizing sodium chlorite, which is a stabilized form of chlorine dioxide. When you dissolve the sodium chlorite in water, it does not generate the chlorine dioxide until it is either acidified or oxidized by an agent such as sodium hypochlorite. Once the chlorine dioxide has formed, it has three times the disinfection capacity of hypochlorite and is not pH sensitive. One of the concerns about the use of chlorine dioxide is the risk of the formation of chlorates and chlorites. There are some concerns with these products and there are considerations being given to them becoming regulated substances. Keep in mind that the disinfections of wells are batch treatments and when the well is flushed would not leave behind these substances.

There are better ways to get disinfection of wells, by considering the concentrations to be used, the contact time, the critical importance of pH limitations, hydraulic aspects, disruption potentials, and the level of protection being afforded to the bacteria by the deposits, tubercles, nodules, scale and biofilm.

Another concern with the use of chlorine-based disinfectants is the potential for the formation of chlorinated organic by-products, such as the trihalomethanes (THMs). If you have some of the organic precursors present in the water and you are using chlorine, then you can expect to get THM formation. We have undertaken a number of studies on wells that we have been chlorinating to examine the THM formation potential. The concern was whether THM molecules were being created above the maximum contaminant level (MCL), and the answer was, yes, during the actual period of disinfecting treatment but not after the wells had been flushed. Well rehabilitation treatments have to be looked at as batch treatments. Once that material had been flushed out of the well after treatment there were no THM molecules detected in the subsequent treated water flows. Even after months, there were still no residual THM molecules detected. This would support the probability that THMs are formed after the shock chlorination has been completed but removed when the well was flushed and redeveloped.

We have used some of the nonhalogenated disinfectants to get away from these risks associated with THM production. There is sensitivity to these risks and we have to have alternate disinfectants available. Another approach used for water treatment is the removal of organic precursors. In Europe, ozonation is often used to reduce the organics to assimilable organic compounds and then biologically removed so that there is a lower potential for THM formation. Using ozone to break down the larger organics into smaller forms reduces the THM risk if they are removed from the water. These smaller organic molecules are now much more easily broken down biologically. This type of approach cannot be considered for wells.

EFFICACY OF USING PELLET CHLORINATORS IN WATER

Another area of concern that really needs to be addressed is the use of pellet chlorinators to suppress the bacteria biofouling in wells and to act as a disinfectant. This method is designed to allow for the continuous chlorination of wells in general.

I am not a great supporter of the use of pellet chlorinators, because when you consider dropping calcium hypochlorite pellets down into a well, I have seen far too many problems resulting. With continuous chlorination I have seen big holes in stainless steel pump jackets because chlorination and stainless steel are not compatible materials. The stainless steel commonly used (Type 304) in the groundwater industry is not selected for its resistance to chlorides, simply because it is not expected to be used in a chloride environment. Chloride-resistant stainless steels do exist and have a lower carbon content (Type 316), but most stainless steels used in the industry are not selected because of a resistance to chloride. Under normal circumstances this would not be expected to be a problem.

In addition to this problem, there can also be calcium deposition potential, depending upon the water chemistry. If they are used in wells that have water chemistry with a high level of bicarbonate, then the calcium from the calcium hypochlorite can form calcium carbonate precipitates. There was an occasion when we had to pull a pump from a well where a pellet chlorinator had been used. The pump was so coated with calcium carbonate that it was almost impossible to pull from the well! The pump, when we finally got it out of the well, had been completely scaled up with a solid block the same diameter as the well casing with these calcium carbonate deposits. I would consider continuous chlorination to be a good mechanism for stopping, or at least slowing down, the so-called iron-related bacteria-type problems. There are, however, too many negative aspects to this form of treatment.

It should be remembered that, under some circumstances, the calcium hypochlorite could become explosive if stored with another strong oxidizing agent. This is because it is a strong oxidizing agent. Again, the calcium hypochlorite (such as HTH) would have to be stored in large quantities to create this risk of explosion. In some of the hazardous well sites where there are free hydrocarbons, this can also create explosions and has to be a concern. We have to, therefore, be cautious at all times, with not just the amounts and concentrations of chemicals being added but also how they can react with other chemicals within that specific environment.

Chemical treatments must be custom-tailored. I cannot over-emphasize the value of that. The importance of deposit analysis needs to be stressed because you are no longer using indirect measures from water chemistry when you actually have an understanding of what your deposits are by doing a deposit analysis. I like to use the analogy: we use pumped water analysis like a doctor would use the blood analysis to determine the problems of a patient. We are doing the same thing, except that instead of the blood we are using a pumped water sample to determine what is wrong with the well. We are using that sample and its analysis to determine what is wrong with our patient (i.e., the well)!

SELECTION OF ACIDS

There are many different acids that can be selected for use in water well rehabilitation (Figure 82). Some of these acids have fairly limited applications. My choice is still hydrochloric acid, also commonly known as muriatic acid.

Sulfamic acid is one major alternative acid. It is fairly widely used in the industry. There is a wide range of products in the market that have sulfamic acid as a major ingredient. Sulfamic acid is fairly good at dissolving calcium carbonate and magnesium carbonate. However, sulfamic acid has almost no ability to dissolve iron and manganese and therefore would not be able to effectively dissolve and remove iron and manganese deposits. Traditionally, this problem has been addressed by the addition of Rochelle salts, for example. This is done to increase the solubility of iron and manganese, but these formulations are too slow and too mild to ever really increase the removal. We do have some chemical blends now, which are combinations of the sulfamic acid with some dispersants such as polymers that can be better at dissolving minerals. Generally, though, if the problem is iron and manganese deposits, there are better chemistries which can be used. Sulfamic acid is chosen most of the time because it is less corrosive. If you do have a calcium carbonate type of scale, the sulfamic acid can be effective. If you have an iron and/or manganese form of scale, it is probably not going to be very effective.

Step 2: Acids

- ❑ **Hydrochloric (muriatic)**
- ❑ **Sulfamic**
- ❑ **Hydroxyacetic**
- ❑ **Citric**
- ❑ **Phosphoric**

Figure 82

Hydroxyacetic acid, or **glycollic acid** as it is sometimes referred to, and citric acid are too mild as organic acids to be very effective at dissolving the often very hard mineral scale. We have recently been involved in trying to combat a problem in a very major city in Texas. This city had written into the specifications for rehabilitation that glycollic acid be used because it is less corrosive (as a replacement for the more corrosive hydrochloric acid). What we demonstrated to them was that it was not the pH of the treatment solution that was so important as the availability of hydrogen ions. It is the hydrogen ions that will ultimately cause the dissolving of the mineral scale. The availability of hydrogen ion determines the strength of an acid.

The technical term for this is the ionization constant. In other words, it reflects the ability of the acid to separate into its ionic species (H^+ and Cl^-). Hydrochloric acid does that fairly readily. Hydroxyacetic acid (glycollic acid) does not do that very readily. Theoretically, this acid is 826,000,000 times slower than hydrochloric acid because of the ionization constant. This means it is going to be less effective because it is not going to make available the hydrogen ions simply because it is not going to separate into the ionic species either fast enough or in enough quantity to dissolve the deposits. Simply having a low pH solution does not mean that you are going to be able to dissolve that mineral scale; there have to be active hydrogen ions present in those treatment solutions. Monitoring pH of the acid in a well is not always an accurate measure of the dissolving potential of an acid or whether the acid is spent. It must be remembered that the pH scale is exponential, and therefore significant differences can exist in the dissolving capacity between a pH of 1 and 2.

This means that the weaker organic acids chosen because of the low risk of corrosion would have much less potential for dissolving mineral scale. They are not really going to be able to achieve dissolution of the mineral scales. These acids work better when they are used in blends, and that is often what we will do. For example, we can mix hydrochloric and glycollic acids with other ingredients and have a good solution to work with. The hydrochloric acid provides a good source of hydrogen ions and the glycollic acid is better as a disinfectant and it is also better at dispersing biological slime. It will be fairly effective at dispersing biological slimes. With this blend, it is the glycollic acid, which provides both the disinfection and slime dispersion functions. It is the hydrochloric acid which is dissolving the mineral deposits.

Phosphoric acid (H_3PO_4) is another acid that is in common use. It has three hydrogen ions that are present but not readily available. The dissociation of the first hydrogen ion is rapid, but the other two dissociate very slowly. This means that the phosphoric acid does not offer the advantages of hydrochloric acid because the dissociation of the hydrogen ions would be too slow. Given the time frame commonly employed in water well rehabilitation (e.g., 12 to 24 hours), phosphoric acid would often be too slow.

Heat can speed up the activities of the chemistries in general in water well rehabilitation. There is a general consideration that for every ten degrees Centigrade rise in temperature the level of chemical activity is doubled. There is no doubt that taking the temperature up would increase the rate of dissolving the mineral deposits, scales and biofilms.

The chemical activity is only one aspect of an effective rehabilitation program. The removal of deposits does not always require increases in temperature.

ADVANCES IN TECHNOLOGIES AND CHEMISTRIES

Over the last decade, there have been considerable advances in the technologies and chemistries for water well rehabilitation. A lot of these improvements in chemical mixtures and formulations were originally developed for scale control in boilers, heat exchangers, and cooling towers (Figure 83). Chemical companies have been developing a whole range of new polymers that have increased the dispersion capacity and penetration capacity of scale and similar deposits.

Formulated Chemistry

Based upon deposit analysis should include:

□ **Dispersants**

□ **Penetrants**

□ **Surfactants**

□ **Corrosion Protection**

Figure 83

We are using these advancements now in the well rehabilitation industry. One of the major advances is the use of "polymeric blends". Some of these blended formulated chemistries that we are now adding to mineral acids include mixtures of polymers. This means that when the traditional chemicals begin to dissolve the deposits, the polymers cause its dispersion for better dissolution of the scale. The major advantage of these large polymeric molecular masses in the treatment solutions is that it prevents the fragmented dissolved and particulate deposit material from recombining. This is effective even over a wide pH range, which gives this form of dispersion a real advantage. The result of this is more material removed when pumping off the chemistry and less of the particulate material that has to be removed from the well during development at the end of the rehabilitation treatment.

This raises another concern that is now being addressed by the use of "polymeric blending". It relates to the ability of the solutions that is being applied or created to carry total dissolved solids (TDS). This is known as the "TDS carrying capacity". If we have hydrochloric acid with a pH of less than 3.0, that solution is capable of carrying approximately 12,000 ppm of TDS.   If the pH now shifts upwards from 3.0 to 4.0, then approximately 80% of the TDS carrying capacity is lost and that solution can now only carry approximately 2,400 ppm of TDS. The remaining solids (above 2,400 ppm) are dropped out of solution when the pH rises from 3.0 to 4.0. This shift is due to radical changes in the charges in the solution, which now allows more of the charged particles to recombine and settle out as precipitates. When polymeric dispersants are added with the hydrochloric acid, the dispersants will prevent the positively and negatively charged materials from recombining and precipitating out. Now it becomes possible, with the addition of the dispersants, to carry up to approximately 120,000 ppm of TDS over the entire pH range. In practice, we have found that raising the pH from 5.0 to 7.0 causes a modest loss of 10% of this TDS carrying capacity. We do still therefore get some TDS dropping out of solution but it clearly is nowhere near as much as if the dispersants had not been used. The bottom line here is the dispersants improve the capacity to carry more minerals in solution or suspension without them precipitating out. Other aspects of rehabilitation are equally if not more important, such as the mechanical agitation, physical disruption, redevelopment and surging of the well. The dispersants have formed that extra step that we need to take in order to ensure more complete deposit removal.

Polymeric and general surfactants (wetting agents) can also be used in rehabilitation. These improve the removal of the materials during treatment. This is because they are very effective at penetrating deposits and scale in general. They therefore help to separate all of that material. For example, if there is a scaled up surface, the application of an acid will cause hydrogen ion activity to work on the surface of the deposit but not deeper down into the deposits. The scale is not porous enough to allow the acid to penetrate. The surfactants and dispersants open up the porosity of the scale and allow the acid (hydrogen ions) to penetrate and dissolve the scale deeper down.

As a practical demonstration of this principle, we started to examine the advantages of using a surfactant and dispersant product approximately six years ago. One of the recommendations given by the manufacturer was to use it by itself. These polymeric dispersion chemistries cannot be used alone. They have to be used in combination with a stronger mineral acid such as hydrochloric. The manufactured product had the polymeric surfactant/dispersants combined with organic acids and corrosion protectant, but it could not deliver the necessary hydrogen ions to dissolve the mineral scale. The first well treated with the surfactant/dispersants product alone had a loss in specific capacity down to approximately 0 gpm/ft! To explore the reasons for this, we went back to the laboratory, taking a piece of scale from that well. When the product was used in the laboratory to treat the scale, the scale swelled up to roughly ten times its original volume. This explained why the well's specific capacity had dropped to zero. The scale had swollen up down-hole and plugged the well shut. Without the acid, the net effect had been that the surfactant/dispersants had entered the scale deposit and opened it up but had no capacity to dissolve the scale. That is, of course, a very beneficial feature but without the acid, the well just seized up! When we went back to that plugged well and treated it with hydrochloric acid as well as the surfactant/dispersants, it dissolved the deposits very effectively and the specific capacity was recovered.

One of these advanced well cleaning chemistries that I have had experience with is QC-21 Well Cleaner™. Incidentally, for those interested, QC-21 stands for quality chemistry of the twenty-first century! It has a good combination of treatment chemicals and includes organic acids, dispersants and surfactants with corrosion protection but has to be added to a stronger mineral acid. It is capable of disinfection and slime dispersion. The reason for using this organic acid in a blend was the treatments that sequenced chlorination with acidization followed by another chlorination. It has become an "industry standard" in that it is a widely practiced technique across the U.S. Essentially the deposit may contain a very high mineral content (normally 75 to 80%), but the majority, if not all, of that material may have been deposited as a result of microbial activity. What the microorganisms are doing is scavenging the organic materials for nutrients and are "casting aside" the mineral salts to then form the scaled up deposits.

The slime is often on the outside of the deposit where the environment is good for their growth. That slime prevents acid contacting the mineral scale and resulting in dissolution. The purpose of the initial chlorination is simply to destroy the integrity of that slime matrix so that the acid, when applied after the first chlorination, can dissolve the exposed mineral structures. The third phase chlorination can now disinfect the surfaces exposed by the removal of the deposits. In other words: kill the slimes, strip the minerals and disinfect the surfaces. QC-21 does the same thing, but in a single stage, by the combination of a disinfecting organic acid (glycollic acid) with a polymeric surfactant as well as passivation of the surface to act as a corrosion protection. When the QC-21 is added to the hydrochloric acid, the first two common steps are reduced to a single phase making for more efficient well treatment. Therefore, the activity of hydrochloric acid can be enhanced by the use of other chemistries appropriate to the forms of problems being experienced in particular wells.

AQUA FREED™

One technique, with which I have had considerable experience, is the use of liquid carbon dioxide applied using the Aqua Freed™ process. This is a patented process for which we have the license in approximately 17 states. This process uses gaseous and liquid carbon dioxide to rehabilitate water wells.

Layne has used the technology on more than 600 wells to-date (Figure 84). These have not been small wells, although the company that we licensed the technology from recently utilized the technique on larger wells after initial use on thousands of smaller private wells, which typically produced 5 to 10 gpm. We have been concentrating more on the municipal and industrial wells with very large production capacities. This process has been used on a wide variety of wells, from shallow to very deep wells, from small diameter to large diameter, etc. Approximately 30% of the wells treated have been "rock" open-hole wells that do not have a well screen but are completed in fractured formations of limestone, dolomite, granite and sandstone. The rest were screened wells, either naturally developed or gravelpacked wells. In fact, the naturally developed, screened wells, which do not have a gravel pack, gave a better success rate than the screened wells with gravel pack. Gravel-packed wells create a more difficult barrier to rehabilitation and a more challenging barrier to redevelopment. This is because there is this additional barrier (i.e., the gravel pack) through which the materials have to migrate during the treatment.

Solution:
Aqua Freed™

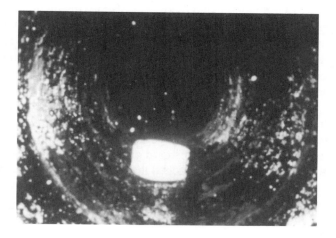

Figure 84

The Aqua Freed™ process is one for which we have kept a very good database on the individual treatments. Our overall success rate been for rock (open-hole) wells, on average, has a return to 110% of the original specific capacities. The equivalent success rate for screened wells has been, on average, returned to 82% of the original specific capacities. This experience has led me to view this technique as being superior to other well rehabilitation procedures. There is no question in my mind that the use of liquid carbon dioxide is a superior technology. I do not say this lightly but I have had the experience that would allow me to back up these statements. In comparative conditions, I have seen this technique outperform many other treatments for restoring lost capacities and solving water quality problems such as high iron or unsafe bacterial samples that have otherwise been difficult to control. These advantages are because this process is more effective at removing deposits from both inside and around the well. This gets back, to one of the fundamental aspects discussed, effective deposit removal.

One major feature that leads me to say this is the excellent penetration that the liquid carbon dioxide achieves into the surrounding formations (Figure 85). We have worked with many wells that were no longer responding to chemical treatments and brought them back. For example, in a large city in Texas, there were twelve wells that were used for aquifer recharge, storage and recovery (ASR). This method involves recharging the aquifer in the winter months and drawing on the aquifer in the summer months by pumping these wells. These wells had developed some very severe plugging problems, and it got to the point that the combination of injection and removal of ground water had driven the plugging zone deeper and deeper into the surrounding formation, beyond the reach of chemical treatments or jetting. When driven this deeply into the formations, chemical or mechanical treatments can no longer reach the plugging zones. We found that, with Aqua Freed^{TM}, these wells were restored close to the original capacities, showing that this treatment had reached these deep plugging formations.

Aqua Freed^{TM}

Benefits:

- **Environmentally Safe**
- **Fast Results**
- **Excellent Penetration**
- **Better Disinfection**

Figure 85

To demonstrate the ability of a treatment using liquid carbon dioxide to penetrate formations, in rock formations, we have had an adjacent well 250' away from the well being treated in which the specific capacity improved from 37 gpm/ft to 94 gpm/ft. That is more than a doubling (154% increase) in the specific capacity of the well that was not even treated!

The reason for this extended treatment zone is that these wells were in a consolidated fractured rock formation and the carbon dioxide was able to extend along these fractures into nearby wells. We have also been able to measure extended penetration benefits in some relief wells installed in sand and gravel formations. Excellent penetration can therefore be seen in screened wells in alluvial formations. Improvements have been seen in wells at distances of up to 30' away from the well being treated. Again, this is because of the penetration capability of the carbon dioxide into the formation. That is one of the primary advantages of this process.

As we continue to use this process, experience has shown that the apparent ability for the Aqua Freed^{TM} process to act as a superior disinfectant may not be justified. This is not only limited to Aqua Freed^{TM} but all rehabilitation procedures, as disinfection is not the key to increasing longevity. When we first started to apply this process, the evidence indicated that we would be able to reverse the trend of shortening time intervals. The short time between treatments was a common occurrence after each traditional rehabilitation treatment. There were occasions at that time when wells which normally lasted only 3 to 4 months between treatments now lasted three to four years after being treated using the Aqua Freed^{TM} process.

At first, I thought that this must be because we were achieving a better disinfection of the well. After all the additional experiences, it is now becoming clear that the Aqua Freed^{TM} process is very effective and there is more complete deposit removal. This means that there is less residual left in the well to start the next round of plugging and, as discussed before, getting to the excess capacity that exists in wells. This means the well is left in a cleaner condition. Once those surfaces have been cleaned you can expect the bacteria to recolonize them in a heart beat, but at least you are starting with clean surfaces since the deposits have been removed. There may be a few days of pumping after the application of carbon dioxide having to pump off carbonated water. The low potential environmental impact of using Aqua Freed^{TM} becomes an attractive feature in its application. One of the big advantages here is that we do not have the problem of disposing of spent chemicals after a treatment. The discharge would still likely have significant particulate material that needs to be disposed of properly.

Depending upon the problem, it is possible to enhance the impact of the Aqua Freed^{TM} treatment by blending other chemicals into the treatment, but most of the time it is used on its own. For example, if there is a bad iron deposit in the well, then additional chemistries will be used to aid in the removal of that iron using the Aqua Freed™ process as the energy of disruption and bringing more iron in solution with chemicals. About 90 to 95% of the time, the Aqua Freed™ process works effectively without additional chemistries. When the wells have been videotaped before and after treatment, the differences are often much better than can be seen with many chemical treatments. Carbon dioxide works effectively as a surface cleaner and this is very evident on the videotapes. It has excellent ability to act as a cleaner, both of the well screens and the wells. In fact, I do have a greater "comfort level" that we can get wells cleaner both visually and with better results in general with carbon dioxide. This is because of the agitation (energy of disruption) it creates, the greater penetration, and the ability to follow the pathways of least resistance.

I have seen more impressive video with this method than various other chemical treatments. Often, with the videos taken after chemical treatments, a lot of deposits can still be seen which can lead people to say, "There is still a lot of stuff in there!" simply because the chemistry has not dissolved and removed all of the "crud". I have observed overall better cleaning when the Aqua Freed™ process has been applied.

We inject carbon dioxide by putting packers just above the top of the well screen (Figure 86) and injecting below the packer. The Aqua Freed™ company itself has treated several thousand small producing wells using this technique. They used cylinder gas for many of these smaller well treatments, while the larger wells require the use of much larger equipment. We inject both gaseous and liquid forms of carbon dioxide into the well and surrounding formation.

The active component from the injected gas is carbonic acid and the lowest pH likely to be achieved is 6.0, and so this is a very mild acid under atmospheric conditions (Figure 87). If, however, there is pressure in an aquifer or a sealed well, then the pH values can become reduced to as low as 5.0. That is still mild. There is a coupled effect of freezing and agitation as the carbon dioxide is injected at approximately $0°$ F. That is the normal temperature for injection based upon the pressure. The coldest temperature that is theoretically achievable is -109°F. The Aqua Freed™ process is not an overly aggressive process and we have very effectively used it without problems on PVC or HDPE wells. We have treated high-density polyethylene wells and horizontal wells without any signs of problems in the wells.

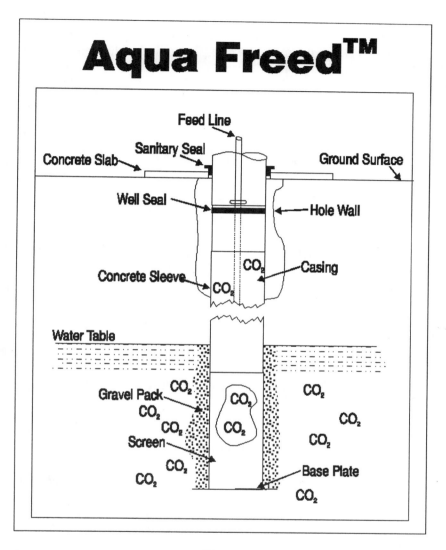

Figure 86

I feel confident the treatment is relatively harmless. This does not mean that the potential for damage to wells does not exist, but that potential exists with any rehabilitation procedure. The freezing activity is localized and I do not believe that all the success is achieved through the freezing but rather because of a combination of carbonic acid, agitation, and freezing. It is not necessary that the whole well be frozen into a solid block of ice in order to get disruption of material! Where the freezing occurs, it is good at slime dispersion, good at agitation, and also good at breaking up mineral scale. This means that the localized freezing does have a positive impact.

Aqua Freed™

Mode of Action:

❑ **Rapid Expansion (500 times)**

❑ **Formation of H_2CO_3**

❑ **Freezing**

❑ **Reduces Surface Tension**

Figure 87

The bulk of the activity is due to the expansion rate from a liquid to a gaseous state of the carbon dioxide. That is an expansion rate of 570 times in volume from the liquid to the gaseous state. The agitation achieved with liquid carbon dioxide is the same action as when dry ice is placed in water. There is a massive foaming as the gases change, in this case, from a solid to a gaseous state. The application of dry ice into wells would not achieve the same benefits because an excessive amount of energy can be placed in the well without being able to relieve it if too much is put in. In addition to not controlling the input of energy, the energy that is put in is not as effectively directed into the surrounding formation with the application of dry ice. Much of that effective energy is lost up the well column. In essence, the Aqua Freed™ process is the controlled injection of carbon dioxide, as pressure on the upper casing, injection pressure and down-hole pressure are monitored to regulate the feed rate that the well will comfortably take. The rate of carbon dioxide feed is regulated to assure that the pressures (and energy) going down the well can dissipate into the surrounding formations.

It is possible using concentrically arrayed tubes to selectively inject into different zones of the well, maximizing the effect of the Aqua Freed^{TM} process (Figure 88). One of our units for supplying carbon dioxide to wells has a vessel with 26-ton storage capacity of liquid carbon dioxide and a vapor space on top, which contains gaseous carbon dioxide (Figure 89).

Figure 88

These tanks are kept at pressures around 330 psi. A vaporizer is used to maintain that pressure by taking the liquid carbon dioxide and converting it to the gaseous form which is injected into the vapor head space of the vessel to maintain head pressures. The vaporizer is also used to inject vapors directly to the well. For wells of less than 700', it is possible to use the ambient pressures in the vessel to force the liquid and gaseous carbon dioxide down the well. Deeper wells require additional pressures. Piston pumps are used to directly pump the liquid carbon dioxide down the well. The deepest well that has been treated to date was 3150' deep.

Figure 89

The original interest in the Aqua Freed™ process was for its application in collector wells consisting of large diameter caissons with laterally projecting screens at the bottom of the caisson (Figure 90). This initial interest was stimulated by research I conducted at the Compagne Generale Des Eaux in France while working on collector wells. I found that there was difficulty in trying to clean those types of wells for which I projected that the Aqua Freed™ process may offer some advantages. I could visualize a diver putting a packer into a lateral (horizontal) well screen and then injecting the carbon dioxide. At that point, once the well screen is isolated, the activity would be the same on a horizontal or vertical screen. A six-day turn-around was found to be achievable on such a radial collector well. Previous techniques could take not just weeks but even months to achieve the same levels of rehabilitation. However, most of the experience with the more than 600 wells has been with vertical wells.

Before

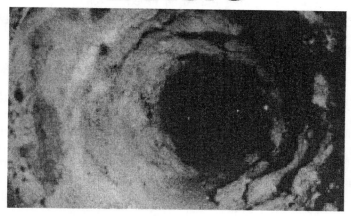

After

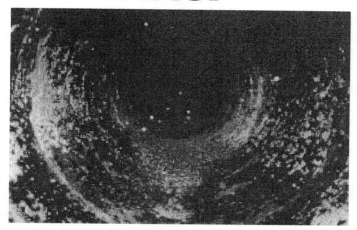

Figure 90

The average recovery in specific capacity in screened wells was 82% of the original capacity, representing a higher average with some which do recover better than original specific capacities. By comparing this treatment to most rehabilitation processes, these results are very good. Again, in limestone and open-hole wells, our average success has been to get the wells back to 110% of original specific capacities. As an example, for an open-hole well in Indiana in a limestone formation with a pretreatment specific capacity of 2.1 gpm/ft and an original specific capacity of 3.9 gpm/ft, after the Aqua Freed™ treatment, the specific capacity increased to 19.5 gpm/ft. This was 5 times better than original. This does indicate that some fracturing of consolidated formations can be achieved. The main application of the process is effective deposit removal, rather than enhancing fracturing. Layne has performed up to a dozen projects on new wells for the purposes of stimulation and enhancing the production. Our overall success rate for those applications also is very good. The costs of Aqua Freed™, incidentally, are comparable to properly applied and selected traditional chemical treatments, and the liquid carbon dioxide can be more cost effective on larger wells. As the size of wells increases, the volumes of chemical required for the surface area in the well increases significantly. By comparing rehabilitation costs on, for example, deep injection wells in Florida, the application of the Aqua Freed™ would be less expensive. In addition to the cost benefit, the potential success with the application of carbon dioxide also exists because of the mode of action already described.

ENVIRONMENTAL SAFETY CONCERNS

This area is now becoming a major concern particularly in the industrial arenas. Essentially, the techniques and chemistries used in water well rehabilitation are really modifications of techniques widely used in the treatment of water and are approved by NSF international. Because of this, I am not overly concerned with the use of chemicals as long as they are handled properly. In addition, the chemicals are not difficult to neutralize or dispose of properly after treatment. That is something that we have had to deal with repeatedly, but it does not create major problems. These are things that have to be dealt with.

Proper handling and proper disposal procedures are essential parts of any water well rehabilitation. Even though this is being done, it has to be remembered that we will not be pumping off clear water during and after treatment. There are going to be muds, dissolved and particulate minerals and biological slimes that will still need to be addressed. The purpose of performing well rehabilitation treatments is to remove this material and therefore will always be dealing with it, regardless of the procedure used to remove it from the well and well environment. Sometimes these can become environmental concerns when addressing the disposal issues. For example, in a large basin in Southern California, the environmental regulations are so strict that we have to treat water on discharge to drinking water standards before it can be discharged. NPDES permits are required in order to effectively control the environmental impact on streams and lakes.

We have to have less than 1.0 NTU of turbidity to remove the dissolved minerals before the water can be discharged into storm channels, etc. This means we have to neutralize, remove excess minerals and filter before it becomes acceptable to allow discharge. We are able to do this on site and can meet any treatment requirement. Obviously these conditions add to the cost of the treatment but most of the time this is not necessary. Most of the time the posttreatment concerns are limited to readjusting the pH towards an acceptable (normally >pH 6) value, dechlorination (if there is residual chlorine), and removal of sludge and iron from the discharge waters during and after treatment. A lot of times this sludge and iron removal can be done simply by "land spreading." It is no different sometimes than the material that you flush from a hydrant onto the street! It is exactly the same material. It is very important to follow federal, state and local regulations.

12

WELL DEVELOPMENT AND MAINTENANCE

WELL DEVELOPMENT

No matter what type of treatment application you are applying to rehabilitate a well, one of the most important steps, if not the most important step, is the development of the well after the treatment has been completed. It is the removal of the dissolved, dislodged and disrupted deposit materials from the well (Figure 91). Once you have removed them from the surfaces it is now necessary to get them out of the well and the surrounding formation. A lot of the hydraulics information already discussed is very relevant in the development phase.

> # Well Development:
>
> ❑ **Involves removal of dislodged or disrupted deposits**
> ❑ **Is one of the most important steps in effective rehabilitation**
> ❑ **Can involve many different techniques**
> ❑ **Must be able to achieve transfer of energy into the surrounding formation**
> ❑ **Must pay "special attention to detail"**

Figure 91

There needs to be a concentration of energy on small sections of the screen in order to force this material out of the well. Effective energy must also be applied into the formations beyond the screen if success is to be assured. Water, clearly, has to be the medium through which the energy is transferred. This energy also cannot be concentrated only in one direction where the material may become packed into the surrounding formation. The water has to be the carrier, regardless of what technique is going to be specifically applied.

Jetting is a commonly used development procedure, but it does not have the penetration capability that was previous thought to occur (Figure 92). Using a physical model, studies have found that either high pressure, air or water, has probable penetration only one or two inches into the surrounding formation. Thus, it is not a very effective means of penetrating into the formations, which is so important in well development. This energy also is concentrated in one direction when jetting is used by itself.

Solution

Mechanical Redevelopment

Mechanical Methods:

- **Brushing**
- **Surge Block**
- **Air Lift Pumping**
- **Jetting**

Figure 92

Swabbing was found to be one of the best procedures. Here, the well is treated with a single agitator or swab in a manner resembling a plunger by moving it up and down. In the physical models, this swabbing has been found to be one of the best development procedures. There is a basic need to surge these wells effectively to drive enough hydraulic "energy" into the formations and to get the detached particulate material moving out of the well and the porous media.

Mechanical redevelopment can involve surging, swabbing and jetting. From my experience, the most effective method has been the use of a combination of airlift and swabbing. By isolating sections of the screen (4', 5' or 10' depending upon the diameter of the well) as zones. Each of these zones can now be developed more effectively. It is the combination of a tool, which moves up and down the zones forcing debris into suspension, then airlifting or pumping debris out of the well from that isolated zone.

Swabbing action and the moving of water creates the velocity here by means of the air or a pump into and out of the formations (Figure 93). This is done in a manner sufficiently aggressive to move the solutions and disrupted particles out of the well.   It can also be looked at as fluidizing the particles by getting enough energy to get the particle

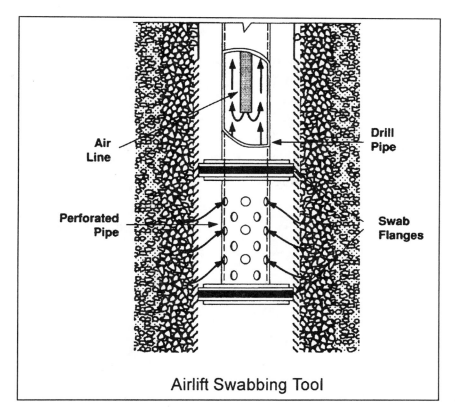

Air
Line

Drill
Pipe

Perforated
Pipe

Swab
Flanges

Airlift Swabbing Tool

Figure 93

moving, and the need to keep it moving out of the well. Research in sediment transport in river bottoms has demonstrated that it takes up to 10 times more energy to get a particle moving than it does to keep it moving.

If the static water level is too low and there is not enough submergence of the airline down the well, there will not be an ability to move the water to carry the energy into the surrounding formation. There has to be an adequate head of water in the well for this practice to be efficient. The velocities will not be there to carry the materials out of the formation, up and out of the well. In these cases (of low head or velocities), a submersible pump may have to be installed to bring up the velocities into the range that will ensure that the energy is achieved and that particulate material is removed from the porous media (Figure 94). Now, with the pump in place, the velocities (pump rate) can be raised into a satisfactory range.

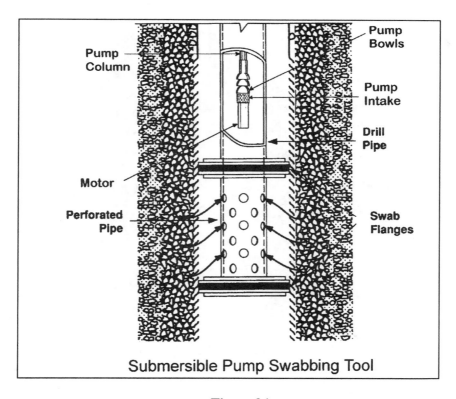

Figure 94

There is a preference for air lift swabbing over submersible pump swabbing if enough submergence exists for the air line. The preference for air lift swabbing is due to the better activity with air lifting over pumping water with a submersible pump. The additional agitation with the surging effect of pumping water with air will aid in development and removal of particles. As previously described, a recent procedure with very good results is the isolation of several zones in the well and circulation through the gravel pack from one small isolated zone and back into the other. The addition of energy achieved with gaseous carbon dioxide during the recirculating process allows more sludge and sediment to be removed.

PREVENTATIVE MAINTENANCE OF WELLS

In the past, a 15 to 20% loss in specific capacity was used to trigger a preventative maintenance treatment. This may already be too late to get the disruptive action with the pump in place because of the potential for deposition to occur prior to impact on the specific capacity. It would be better to use a time frame approach (e.g., every three months or six months or once a year) than to use the losses in a specific capacity and it is clearly better to do it more frequently rather than less. The time frame between these scheduled approaches would be determined geo-graphically from experience in an area, well field or well. Where a well has excess production capacity, then the specific capacity may not be a true measure of hydraulic efficiency. In these cases, you could actually loose a lot of the porous media capacity, much of the aquifer capacity or production capacity. A lot of the porous media surrounding the well could be plugged without impacting on the specific capacity. In these cases where a well has excess capacity, then the preventative maintenance needs to be performed as soon as there are any losses in specific capacity. That can be fairly frequent from time to time.

I do not recommend shock chlorination as preventative maintenance chemistry. I have already mentioned that shock chlorination is too limited in its ability to remove deposited material. We start losing the effectiveness of shock chlorination once the microbes have developed their protective mechanisms that are deliberately put in place by the bacteria. Those protective mechanisms are basically the biofilms with associated mineral deposits. Once they have formed these protective mechanisms, we are often not going to be able to impact them with only disinfectants and possibly with other chemicals.

We are better off looking at more broad-range chemistries. One example of broad-range chemistries would be to use some hydrochloric acid as a part of that treatment with the QC-21 Well Cleaner™. This is to aid in the removal of the deposits that will just be starting to build up and can therefore still be removed with the pump in place. The mineral deposits associated with the bacteria would not be removed by shock chlorination but would be removed by the formulated chemical treatment and so expose more of the biofilms to the treatment. More complete removal becomes achievable when the acid (preferably hydrochloric) is mixed with polymeric dispersants (QC-21 Well Cleaner™). You can start to get more complete removal of those deposits just as they are starting to build up. If you do that frequently enough you can still get the action necessary with the pump in place and you can slow down the amount of significant plugging of the well.

There are problems, which can affect preventative maintenance treatments, for example, seasonal wells or wells that are taken off-line for a period of time. If a well is taken off-line for any period of time, the well needs to be exercised (Figure 95). If the well has been off-line (inactive) for six years, the general attitude is, don't worry about it, but do a bacterial test, inspection, and possibly disinfection before putting the well back on-line. When a well is off-line because of low demand and is needed immediately when the demand increases, without an inspection or testing, there needs to be a preventative maintenance program. These wells do need to be exercised periodically. That exercise should be every three to four weeks and involve between two hours and one day of pumping. The idea behind this is that you want to

Maintenance of Wells

❑ **Preventative Maintenance**

❑ **Operating Schedule for Idle Wells**

❑ **Operating Schedule for Seasonal Wells**

Figure 95

prevent the establishment of an anaerobic (anoxic) condition in the well. When we are operating a well, we are essentially creating an active oxidized zone in the well through the pumping activity. The aerobic zones are being created for a certain distance into the porous media around the well. When a well is shut down for any length of time, all of those oxidized, deposited species begin to be transformed, predominantly biologically, to reduced species in the now anaerobic zone. This anaerobic zone will create water that is often very smelly (often with the "rotten egg" odor), cloudy or black. The simple act of running the pump will prevent the creation of that anaerobic front in the well and stabilize the oxidized species around the well. All on-demand wells should be operated (exercised) periodically to control this anaerobic fouling.

LAYNE ANOXIC BLOCK SYSTEM (LABS™)

Another system that I have had personal experience with is the patented Layne Anoxic Block System (LABS™) that utilizes an inert gas to prevent oxygen from going down the well (Figure 96).

Layne Anoxic Block System
LABS™ - Patented:

❑ **Slows the rate of lost capacity**
❑ **Can effectively control water quality problems**
❑ **Can obtain and maintain bacterial "safe" samples**

Figure 96

There are about 50 of these units installed across the U.S. for various problems such as rapid losses in specific capacity. Some of these wells had become operational nightmares prior to the installation of LABS™, simply because of plugging problems. These plugging problems can be as short as every two to three weeks which will, literally, stop pumping water. Usually these wells are in the environmental industry and have extraction rates ranging from approximately ten to twenty five gpm, which are not high. They stop pumping because the organic contaminants feed the bacteria very rapidly in these wells, causing aerobically driven plugging. The anoxic block restricts the entry of oxygen and so reduces the amount of aerobic activity causing plugging.

The concept here is very simple, based upon nitrogen being used as the inert gas to control aerobic microbial growth (Figure 97). This is not the only place that nitrogen is used in this way. Nitrogen replacement of oxygen is also used widely in the food industry and also in the control of aerobic fouling in water storage tanks.

Solution:

Layne Anoxic Block System™

Properties:

- ❑ **Prevent Penetration of Atmospheric O₂**
- ❑ **Inhibits Aerobic Bacterial Growth**

Benefits

- ❑ **Slows Plugging of Wells**
- ❑ **Prevents and Solves Water Quality Problems**

Figure 97

The Layne Anoxic Block System is based upon the concept that most of the problems in water wells are caused biologically. Most of those bacteria causing these problems are aerobic in nature. The anaerobic bacteria do not cause as many deposition problems. Most of the oxygen enters the well down the well column itself rather than from the points of recharge. The objective is therefore to stop the oxygen from coming down the well column. On some occasions, installing an anoxic block has allowed after several weeks safe samples to be collected from a well that had previously given unsafe samples for years after many attempts at rehabilitation treatments. This is due to controlling the growth of aerobic bacteria that had generated the unsafe sample conditions. There was another application of the Layne Anoxic Block System for two wells in Michigan that had a *Pseudomonas aeruginosa* problem that could not be corrected with disinfection treatments. A few weeks after the installation of the Anoxic Block on these wells, the *Pseudomonas aeruginosa* was no longer detected in the weekly sampling and enumeration.

If using the Anoxic Block for total coliform problems, a high nitrate concentration needs to be considered. When oxygen is excluded from the well water column, some bacteria revert to other alternatives to oxygen. One of these alternate respiratory substrates (facultative anaerobes) is nitrate. For example, some of the total coliform bacteria are able to respire, using nitrate under these artificially created anaerobic conditions. As the bacteria respire using the nitrate, it is converted to nitrite and ultimately to nitrogen gas. Even though good successes have been obtained using the anoxic block for total coliform problems, it would be important to determine the alternate electron acceptors.

One of the concerns often addressed with the use of the patented nitrogen-induced anoxic block is the economics. There is an ongoing cost to the operation of that block. If, for example, you know you can get a twenty year operating life out of a well anyway, you are not going to want to maintain an anoxic block with nitrogen gas for that length of time on the well. That would be another maintenance item to take care of. On the other hand, if you have wells that are experiencing problems every year, six months, or less, and the cost to clean that well is $5,000 to $15,000, this can become an attractive economical alternative.

Most of the time nitrogen is supplied in cylinders to the customers using the Anoxic Block System, but there are a couple of clients where nitrogen generators are used (Figure 98). These self-contained systems involve the extraction of nitrogen from the air using generators which produce 99.9% pure nitrogen gas.

Chlorine chemistries, when used for preventative maintenance, require considerations. The first consideration that needs to be applied to chlorine is that its chemistries are limited at the best of times. Chlorine chemistry has been used too much, not from the point of view of health safety (i.e., disinfection), as a general well cleaner when it is actually too limited in its real effectiveness. While it is a strong oxidizing agent, it is not going to be effective for deposit removal, which is the source of most of the problems.

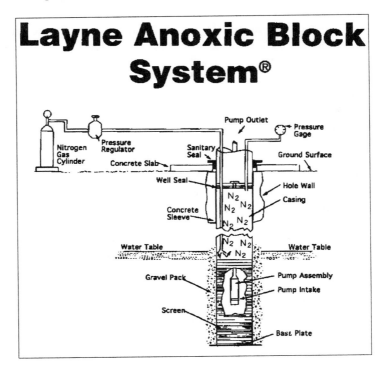

Figure 98

CLOSING REMARKS

I would like to thank the Regina Water Research Institute of the University of Regina for this opportunity to present some of the experiences that I have had in water well rehabilitation. Hopefully this information will prove useful to you. There are some radical concepts that I presented simply because the literature is dated and lags well behind field experiences. Unfortunately, because of this lag we do not have good literature to refer to and much of what I have been presenting is contrary to the literature that does exist! The current initiative by the American Water Works Association Research Foundation on water well rehabilitation is a welcome step to begin to improve the basis for understanding all of the aspects of water well rehabilitation.

If the industry is going to meet some of the regulations put on water quality, such as "unsafe" bacterial samples, we need to take a more aggressive approach to cleaning wells. In addition to the more aggressive approach we need to become more diligent in our control of deposits on surfaces in water environments by preventative maintenance applications.

SELECTED BIBLIOGRAPHY

Biological Aspects of Groundwater at NGWA's 1997 National Convention and Exposition. Las Vegas, Nevada. September 3-6, 1997.

Filermans, C.B. and Hazen, T.C. (Eds.), *Proceedings of The First International Symposium On Microbiology Of The Deep Subsurface.* Orlando, Fl. WSRC Information Services, Aiken, SC. January 15-19, 1990.

Fletcher, M. Bacterial Colonization of Solid Surfaces In Subsurface Environments, pp. 7-3 to 7-10. In C.B. Filermans and T.C. Hazen (Eds.), *Proceedings of The First International Symposium On Microbiology Of The Deep Subsurface.* January 15-19, 1990, Orlando, Fl. WSRC Information Services, Aiken, SC, 1991.

Henderson, A. *A Survey of The Quality of Water Drawn From Domestic Wells In Nine Midwestern States.* Centers for Disease Control and Prevention. (In press). 1998.

Kranowski, K.M., Sinn, C.A., and Balkwill, D.L. Attached and Unattached Bacterial Populations in Deep Aquifer Sediments from a Site in South Carolina, pp. 5-25 to 5-29. In C.B. Filermans and T.C. Hazen (Eds.), *Proceedings of The First International Symposium On Microbiology of the Deep Subsurface.* January 15-19, 1990, Orlando, Fl. WSRC Information Services, Aiken, SC. 1991.

Mansuy, N. *Model Well Systems for Observing The Generation and Control of Biofouling.* Masters of Science Thesis. University of Regina, Regina, Saskatchewan, Canada, 1988.

Mansuy, N., Nuzman, C., and Cullimore, D.R. Well Problem Identification and Its Importance in Well Rehabilitation. Water Wells Monitoring, Maintenance, Rehabilitaiton. *Proceedings of the International Groundwater Engineering Conference* held at Cranfield Institute of Technology, UK. September 6-8, 1990. P. Howsam, E. and F.N. Spon (Eds.),1990.

Web Sites:
http://www.sciam.com/1096issue/1096onstott.html - 9/13/97
http://bordeaux.uwaterloo.ca/biol447new/subsurfacemicrobiology.htm- 9/13/97
http://www.easynet.on.ca/`pic/facts/fact37.htm - 9/15/97
http://www.atlas.co.uk/listons/analysis/fr0132.htm -9/15/97
http://www.igsb.uiowa.edu/htmls/pubs/abstract/tis19.htm - 9/8/97
http://www.dbi.sk.ca – 11/6/98

A

B

blue-green algae · 54
bridge slot screen. · 5
brushing · 152

C

calcium carbonate · 89
calcium hypochlorite · 127, 128, 129, 131, 132
causes of well plugging · 83
cemented gravel pack · 6
chemical treatment · 115
chemoorganoheterotrophs · 25
chloride · 48, 132
chlorination · 48, 51, 53, 60, 86, 115, 121, 127, 129, 131, 133, 138, 149
chlorine dioxide · 128
chlorine gas · 127, 128
chlorine treatments · 61
citric · 125, 133
Citrobacter · 25, 93, 109
Colilert™ · 60, 61
Colisure™ · 60
collapse strength · 5
collector wells · 145
color problems · 52
consolidated formations · 3
corrosion inhibition · 125
corrosion problems · 44, 47
corrosion protection · 136
cost-benefit analysis · 39
costs of well rehabilitation · 37
Crenothrix · 91
CRUD · 32
Cryptosporidium · 28
cycles of fouling · 30

D

dead ends · 65, 67
degradation · 27, 48
denitrification · 76, 78, 79, 80
deposit analysis · 102, 132, 136
deposit problems · 89
depth · 8, 12, 21, 65, 66, 67, 68, 71, 74
detached debris · 72
detachment · 20, 24, 48, 49, 51, 55, 56, 58, 59, 60, 64, 83, 108, 110, 117, 120
detachment of bacteria · 20, 109
deterioration of water quality · 23, 24
developing gravel pack wells · 10

E

F

G

H

I

N

O

P

Q

R

S

T

U

V

W

Z